ARMS AND INFLUENCE

Written under the auspices of the
Center for International Affairs
Harvard University

Delivered in part as the
Henry L. Stimson Lectures
Yale University

ARMS AND INFLUENCE

BY THOMAS C. SCHELLING

NEW HAVEN AND LONDON
YALE UNIVERSITY PRESS

Designed by Sally Sullivan.
Set in Times Roman type,
and printed in the United States of America

Library of Congress catalog card number: 66–15744
ISBN: 978–0–300–00221–1–8

32   31

One of the lamentable principles of human productivity is that it is easier to destroy than to create. A house that takes several man-years to build can be burned in an hour by any young delinquent who has the price of a box of matches. Poisoning dogs is cheaper than raising them. And a country can destroy more with twenty billion dollars of nuclear armament than it can create with twenty billion dollars of foreign investment. The harm that people can do, or that nations can do, is impressive. And it is often used to impress.

The power to hurt—the sheer unacquisitive, unproductive power to destroy things that somebody treasures, to inflict pain and grief—is a kind of bargaining power, not easy to use but used often. In the underworld it is the basis for blackmail, extortion, and kidnapping, in the commercial world for boycotts, strikes, and lockouts. In some countries it is regularly used to coerce voters, bureaucrats, even the police; and it underlies the humane as well as the corporal punishments that society uses to deter crime and delinquency. It has its nonviolent forms like the sit-ins that cause nuisance or loss of income, and its subtle forms like the self-inflicted violence that sheds guilt or shame on others. Even the law itself can be exploited: since the days of early Athens, people have threatened lawsuits to extort money, owed them or not. It is often the basis for discipline, civilian and military; and gods use it to exact obedience.

The bargaining power that comes from the physical harm a nation can do to another nation is reflected in notions like deterrence, retaliation and reprisal, terrorism and wars of nerve, nuclear blackmail, armistice and surrender, as well as in reciprocal efforts to restrain that harm in the treatment of prisoners, in the limitation of war, and in the regulation of armaments. Military force can sometimes be used to achieve an objective forcibly, without persuasion or intimidation; usually, though—throughout history but particularly now— military potential is used to influence other countries, their government or their people, by the harm it could do to them. It may be used skillfully or clumsily, and it can be used for evil or in self protection, even in the pursuit of peace; but used as bargaining power it is part of diplomacy—the uglier, more negative, less civilized part of diplomacy—nevertheless, diplomacy.

There is no traditional · name for this kind of diplomacy. It is not "military strategy," which has usually meant the art or science of military victory; and while the object of victory has traditionally been described as "imposing one's will on the enemy," how to do *that* has typically received less attention than the conduct of campaigns and wars. It is a part of diplomacy that, at least in this country, was abnormal and episodic, not central and continuous, and that was often abdicated to the military when war was imminent or in progress. For the last two decades, though, this part of diplomacy has been central and continuous; in the United States there has been a revolution in the relation of military to foreign policy at the same time as the revolution in explosive power.

I have tried in this book to discern a few of the principles that underlie this diplomacy of violence. "Principles" may be too pretentious a term, but my interest has been in how countries do use their capacity for violence as bargaining power, or at least how they try to use it, what the difficulties and dangers are and some of the causes of success or failure. Success to some extent, failure even more, is not an exclusively competitive notion; when violence is involved, the interests even of adversaries overlap. Without the overlap there would be no bargaining, just a tug-of-war.

But this is not a book about policy. I have not tried to re-organize NATO, to contain Communist China, to liberate Cuba, to immobilize the Vietcong, or to keep India from wanting nuclear weapons; I have not tried to support or to depreciate the manned bomber, nuclear-powered vessels, or ballistic-missile defenses; I have not tried to choose between death and surrender, or to reorganize the armed services. Principles rarely lead straight to policies; policies depend on values and purposes, predictions and estimates, and must usually reflect the relative weight of conflicting principles. (Policies ought to be consistent, but interesting principles almost always conflict.) At the same time, I am sure I have not evenly disguised my prejudices; at some points they obtrude blatantly, at some the reader may share them and not notice them, and at some points I have undoubtedly bent so far backward as to be credited with views I do not hold.

I have not included much on arms control. I wrote, with Morton H. Halperin, a little book on arms control in 1960–61; I still like it and saw no reason either to repeat it or to rewrite it here. There is little on insurgency, revolt, or domestic terrorism; that will have to be another book. There is little or nothing specifically on a "depolarized" world of many nuclear powers, although what I have written, if valid in a polarized world, is probably just as valid in a world of several competing powers, as pertinent to French or Chinese policy as to American or Soviet. And if what I have said is pertinent to the present, it ought to be pertinent in the future, just more incomplete.

I have used some historical examples, but usually as illustration, not evidence. For browsing in search of ideas, Caesar's *Conquest of Gaul* is rich reading and Thucydides' *Peloponnesian War* the best there is, whatever their historical merits—even if read as pure fiction. I have often used recent examples to illustrate some point or tactic; mention does not mean approval, even where a policy was successful. The several pages examining the 1964 bombing in the Gulf of Tonkin do not mean that I approve of it (though, in fact, I do); the several pages on coercive aspects of the bombing of North Vietnam in 1965 do not mean I approve of it (and, in fact, I am not sure yet); the several pages on the tactic of cultivating irrationality at the highest level of government, to make otherwise incredible threats sound credible, do not mean that I approve of it (and, in fact, I do not).

I have had so much help in writing this book that I am tempted to break the rules and let others share the blame as well as any credit. Forceful critics have had a good deal of influence on its shape and style. Two of them, Bernard Brodie and James E. King, Jr., out of great dissatisfaction with the manuscript and even greater affection for its author, took extraordinary pains with every chapter and I must not only thank them here but record that they remain dissatisfied. Others who unhesitatingly told me where I was wrong, or my language unclear, or my book badly structured, or who added ideas and lent me examples were Robert R. Bowie, Donald S. Bussey, Lincoln P. Bloomfield, Thomas C. Donahue, Robert Erwin, Lawrence S. Finkelstein, Roger Fisher, Robert N. Ginsburgh, Morton H. Halperin, Fred C. Iklé, William W. Kaufmann, Henry A. Kissinger, Robert A. Levine, Nathan Leites, Jesse Orlansky, George H. Quester, and Thomas W. Wolfe. This enumeration does not absolve still others from influence over the character and contents of the book.

I have integrated into some of the chapters, in revised form, parts of some articles that were earlier published by the *Bulletin of the Atomic Scientists, Foreign Affairs, The Virginia Quarterly Review, World Politics,* the Center for International Studies, Princeton University, the Center for Strategic Studies, Georgetown University, the Institute of International Studies, University of California, Berkeley, and the Institute for Strategic Studies, London. I appreciate their permission to do this.

A loyal group at the Institute for Defense Analyses in Washington spent eleven weekly seminars with me while the first draft was taking shape. The final draft was put together while I was a guest at the Institute for Strategic Studies in London.

In the spring of 1965 my former colleagues at Yale University invited me to inaugurate, with lectures drawn from this book, the Henry L. Stimson Lectures.

T. C. S.

Cambridge, Massachusetts
November 15, 1965

CONTENTS

1
THE DIPLOMACY
OF VIOLENCE

The usual distinction between diplomacy and force is not merely in the instruments, words or bullets, but in the relation between adversaries—in the interplay of motives and the role of communication, understandings, compromise, and restraint. Diplomacy is bargaining; it seeks outcomes that, though not ideal for either party, are better for both than some of the alternatives. In diplomacy each party somewhat controls what the other wants, and can get more by compromise, exchange, or collaboration than by taking things in his own hands and ignoring the other's wishes. The bargaining can be polite or rude, entail threats as well as offers, assume a status quo or ignore all rights and privileges, and assume mistrust rather than trust. But whether polite or impolite, constructive or aggressive, respectful or vicious, whether it occurs among friends or antagonists and whether or not there is a basis for trust and goodwill, there must be some common interest, if only in the avoidance of mutual damage, and an awareness of the need to make the other party prefer an outcome acceptable to oneself.

With enough military force a country may not need to bargain. Some things a country wants it can take, and some things it has it can keep, by sheer strength, skill and ingenuity. It can do this *forcibly,* accommodating only to opposing strength, skill, and ingenuity and without trying to appeal to an enemy's wishes. Forcibly a country can repel and expel, penetrate and occupy, seize, exterminate, disarm and disable, confine, deny access, and directly frustrate intrusion or attack. It can, that is, if it has enough strength. "Enough" depends on how much an opponent has.

There is something else, though, that force can do. It is less military, less heroic, less impersonal, and less unilateral; it is uglier, and has received less attention in Western military strategy. In addition to seizing and holding, disarming and confining, penetrating and obstructing, and all that, military force can be used *to hurt*. In addition to taking and protecting things of value it can *destroy* value. In addition to weakening an enemy militarily it can cause an enemy plain suffering.

Pain and shock, loss and grief, privation and horror are always in some degree, sometimes in terrible degree, among the results of warfare; but in traditional military science they are incidental, they are not the object. If violence can be done incidentally, though, it can also be done purposely. The power to hurt can be counted among the most impressive attributes of military force.

Hurting, unlike forcible seizure or self-defense, is not unconcerned with the interest of others. It is measured in the suffering it can cause and the victims' motivation to avoid it. Forcible action will work against weeds or floods as well as against armies, but suffering requires a victim that can feel pain or has something to lose. To inflict suffering gains nothing and saves nothing directly; it can only make people behave to avoid it. The only purpose, unless sport or revenge, must be to influence somebody's behavior, to coerce his decision or choice. To be coercive, violence has to be anticipated. And it has to be avoidable by accommodation. The power to hurt is bargaining power. To exploit it is diplomacy—vicious diplomacy, but diplomacy.

The Contrast of Brute Force with Coercion

There is a difference between taking what you want and making someone give it to you, between fending off assault and making someone afraid to assault you, between holding what people are trying to take and making them afraid to take it, between losing what someone can forcibly take and giving it up to avoid risk or damage. It is the difference between defense and deterrence, between brute force and intimidation, between conquest and blackmail, between action and threats. It is the difference between the unilateral, "undiplomatic" recourse to strength, and coercive diplomacy based on the power to hurt.

The contrasts are several. The purely "military" or "undiplomatic" recourse to forcible action is concerned with enemy strength, not enemy interests; the coercive use of the power to hurt, though, is the very exploitation of enemy wants and fears. And brute strength is usually measured relative to enemy strength, the one directly opposing the other, while the power to hurt is typically not reduced by the enemy's power to hurt in return. Opposing strengths may cancel each other, pain and grief do not. The willingness to hurt, the credibility of a threat, and the ability to exploit the power to hurt will indeed depend on how much the adversary can hurt in return; but there is little or nothing about an adversary's pain or grief that directly reduces one's own. Two sides cannot both overcome each other with superior strength; they may both be able to hurt each other. With strength they can dispute objects of value; with sheer violence they can destroy them.

And brute force succeeds when it is used, whereas the power to hurt is most successful when held in reserve. It is the *threat* of damage, or of more damage to come, that can make someone yield or comply. It is *latent* violence that can influence someone's choice—violence that can still be withheld or inflicted, or that a victim believes can be withheld or inflicted. The threat of pain tries to structure someone's motives, while brute force tries to overcome his strength. Unhappily, the power to hurt is often communicated by some performance of it. Whether it is sheer terroristic violence to induce an irrational response, or cool premeditated violence to persuade somebody that you mean it and may do it again, it is not the pain and damage itself but its influence on somebody's behavior that matters. It is the expectation of *more* violence that gets the wanted behavior, if the power to hurt can get it at all.

To exploit a capacity for hurting and inflicting damage one needs to know what an adversary treasures and what scares him and one needs the adversary to understand what behavior of his will cause the violence to be inflicted and what will cause it to be withheld. The victim has to know what is wanted, and he may have to be assured of what is not wanted. The pain and suffering have to appear *contingent* on his behavior; it is not alone the threat that is effective—the threat of pain or loss if he fails to comply—but the corresponding assurance, possibly an implicit one, that he can avoid the pain or loss if he does comply. The prospect of certain death may stun him, but it gives him no choice.

Coercion by threat of damage also requires that our interests and our opponent's not be absolutely opposed. If his pain were our greatest delight and our satisfaction his greatest woe, we would just proceed to hurt and to frustrate each other. It is when his pain gives us little or no satisfaction compared with what he can do for us, and the action or inaction that satisfies us costs him less than the pain we can cause, that there is room for coercion. Coercion requires finding a bargain, arranging for him to be better off doing what we want—worse off not doing what we want—when he takes the threatened penalty into account.

It is this capacity for pure damage, pure violence, that is usually associated with the most vicious labor disputes, with racial disorders, with civil uprisings and their suppression, with racketeering. It is also the power to hurt rather than brute force that we use in dealing with criminals; we hurt them afterward, or threaten to, for their misdeeds rather than protect ourselves with cordons of electric wires, masonry walls, and armed guards. Jail, of course, can be either forcible restraint or threatened privation; if the object is to keep criminals out of mischief by confinement, success is measured by how many of them are gotten behind bars, but if the object is to *threaten* privation, success will be measured by how few have to be put behind bars and success then depends on the subject's understanding of the consequences. Pure damage is what a car threatens when it tries to hog the road or to keep its rightful share, or to go first through an intersection. A tank or a bulldozer can force its way regardless of others' wishes; the rest of us have to threaten damage, usually mutual damage, hoping the other driver values his car or his limbs enough to give way, hoping he sees us, and hoping he is in control of his own car. The threat of pure damage will not work against an unmanned vehicle.

This difference between coercion and brute force is as often in the intent as in the instrument. To hunt down Comanches and to exterminate them was brute force; to raid their villages to make them behave was coercive diplomacy, based on the power to hurt. The pain and loss to the Indians might have looked much the same one way as the other; the difference was one of purpose and effect. If Indians were killed because they were in the way, or somebody wanted their land, or the authorities despaired of making them behave and could not confine them and decided to exterminate them, that was pure unilateral force. If *some* Indians were killed to make *other* Indians behave, that was coercive violence—or intended to be, whether or not it was effective. The Germans at Verdun perceived themselves to be chewing up hundreds of thousands of French soldiers in a gruesome "meatgrinder." If the purpose was to eliminate a military obstacle—the French infantryman, viewed as a military "asset" rather than as a warm human being—the offensive at Verdun was a unilateral exercise of military force. If instead the object was to make the loss of young men—not of impersonal "effectives," but of sons, husbands, fathers, and the pride of French manhood—so anguishing as to be unendurable, to make surrender a welcome relief and to spoil the foretaste of an Allied victory, then it was an exercise in coercion, in applied violence, intended to offer relief upon accommodation. And of course, since any use of force tends to be brutal, thoughtless, vengeful, or plain obstinate, the motives themselves can be mixed and confused. The fact that heroism and brutality can be either coercive diplomacy or a contest in pure strength does not promise that the distinction will be made, and the strategies enlightened by the distinction, every time some vicious enterprise gets launched.

The contrast between brute force and coercion is illustrated by two alternative strategies attributed to Genghis Khan. Early in his career he pursued the war creed of the Mongols: the vanquished can never be the friends of the victors, their death is

necessary for the victor's safety. This was the unilateral extermination of a menace or a liability. The turning point of his
career, according to Lynn Montross, came later when he discovered how to use his power to hurt for diplomatic ends. "The
great Khan, who was not inhibited by the usual mercies, conceived the plan of forcing captives—women, children, aged
fathers, favorite sons—to march ahead of his army as the first
potential victims of resistance."[1] Live captives have often
proved more valuable than enemy dead; and the technique discovered by the Khan in his maturity remains contemporary.
North Koreans and Chinese were reported to have quartered
prisoners of war near strategic targets to inhibit bombing attacks by United Nations aircraft. Hostages represent the power
to hurt in its purest form.

Coercive Violence in Warfare

This distinction between the power to hurt and the power to
seize or hold forcibly is important in modern war, both big war
and little war, hypothetical war and real war. For many years
the Greeks and the Turks on Cyprus could hurt each other indefinitely but neither could quite take or hold forcibly what they
wanted or protect themselves from violence by physical means.
The Jews in Palestine could not expel the British in the late
1940s but they could cause pain and fear and frustration
through terrorism, and eventually influence somebody's decision. The brutal war in Algeria was more a contest in pure
violence than in military strength; the question was who would
first find the pain and degradation unendurable. The French
troops preferred—indeed they continually tried—to make it a
contest of strength, to pit military force against the nationalists'
capacity for terror, to exterminate or disable the nationalists
and to screen off the nationalists from the victims of their violence. But because in civil war terrorists commonly have access
to victims by sheer physical propinquity, the victims and their
properties could not be forcibly defended and in the end the

1. Lynn Montross, *War Through the Ages* (3d ed. New York, Harper and Brothers,
1960), p. 146.

French troops themselves resorted, unsuccessfully, to a war of pain.

Nobody believes that the Russians can take Hawaii from us, or New York, or Chicago, but nobody doubts that they might destroy people and buildings in Hawaii, Chicago, or New York. Whether the Russians can conquer West Germany in any meaningful sense is questionable; whether they can hurt it terribly is not doubted. That the United States can destroy a large part of Russia is universally taken for granted; that the United States can keep from being badly hurt, even devastated, in return, or can keep Western Europe from being devastated while itself destroying Russia, is at best arguable; and it is virtually out of the question that we could conquer Russia territorially and use its economic assets unless it were by threatening disaster and inducing compliance. It is the power to hurt, not military strength in the traditional sense, that inheres in our most impressive military capabilities at the present time. We have a Department of *Defense* but emphasize *retaliation*—"to return evil for evil" (synonyms: requital, reprisal, revenge, vengeance, retribution). And it is pain and violence, not force in the traditional sense, that inheres also in some of the least impressive military capabilities of the present time—the plastic bomb, the terrorist's bullet, the burnt crops, and the tortured farmer.

War appears to be, or threatens to be, not so much a contest of strength as one of endurance, nerve, obstinacy, and pain. It appears to be, and threatens to be, not so much a contest of military strength as a bargaining process—dirty, extortionate, and often quite reluctant bargaining on one side or both—nevertheless a bargaining process.

The difference cannot quite be expressed as one between the *use* of force and the *threat* of force. The actions involved in forcible accomplishment, on the one hand, and in fulfilling a threat, on the other, can be quite different. Sometimes the most effective direct action inflicts enough cost or pain on the enemy to serve as a threat, sometimes not. The United States threatens the Soviet Union with virtual destruction of its society in the event of a surprise attack on the United States; a hundred million deaths are awesome as pure damage, but they are useless in stopping the Soviet attack—especially if the threat is to do it all afterward anyway. So it is worthwhile to keep the concepts distinct—to distinguish forcible action from the threat of pain —recognizing that some actions serve as both a means of forcible accomplishment and a means of inflicting pure damage, some do not. Hostages tend to entail almost pure pain and damage, as do all forms of reprisal after the fact. Some modes of self-defense may exact so little in blood or treasure as to entail negligible violence; and some forcible actions entail so much violence that their threat can be effective by itself.

The power to hurt, though it can usually accomplish nothing directly, is potentially more versatile than a straightforward capacity for forcible accomplishment. By force alone we cannot even lead a horse to water—we have to drag him—much less make him drink. Any affirmative action, any collaboration, almost anything but physical exclusion, expulsion, or extermination, requires that an opponent or a victim *do* something, even if only to stop or get out. The threat of pain and damage may make him want to do it, and anything he can do is potentially susceptible to inducement. Brute force can only accomplish what requires no collaboration. The principle is illustrated by a technique of unarmed combat: one can disable a man by various stunning, fracturing, or killing blows, but to take him to jail one has to exploit the man's own efforts. "Come-along" holds are those that threaten pain or disablement, giving relief as long as the victim complies, giving him the option of using his own legs to get to jail.

We have to keep in mind, though, that what is pure pain, or the threat of it, at one level of decision can be equivalent to brute force at another level. Churchill was worried, during the early bombing raids on London in 1940, that Londoners might panic. Against people the bombs were pure violence, to induce their undisciplined evasion; to Churchill and the government, the bombs were a cause of inefficiency, whether they spoiled transport and made people late to work or scared people and made them afraid to work. Churchill's decisions were not going to be coerced by the fear of a few casualties. Similarly on the battlefield: tactics that frighten soldiers so that they run, duck their heads, or lay down their arms and surrender represent coercion based on the power to hurt; to the top command, which is frustrated but not coerced, such tactics are part of the contest in military discipline and strength.

The fact that violence—pure pain and damage—can be used or threatened to coerce and to deter, to intimidate and to blackmail, to demoralize and to paralyze, in a conscious process of dirty bargaining, does not by any means imply that violence is not often wanton and meaningless or, even when purposive, in danger of getting out of hand. Ancient wars were often quite "total" for the loser, the men being put to death, the women sold as slaves, the boys castrated, the cattle slaughtered, and the buildings leveled, for the sake of revenge, justice, personal gain, or merely custom. If an enemy bombs a city, by design or by carelessness, we usually bomb his if we can. In the excitement and fatigue of warfare, revenge is one of the few satisfactions that can be savored; and justice can often be construed to demand the enemy's punishment, even if it is delivered with more enthusiasm than justice requires. When Jerusalem fell to the Crusaders in 1099 the ensuing slaughter was one of the bloodiest in military chronicles. "The men of the West literally waded in gore, their march to the church of the Holy Sepulcher being gruesomely likened to 'treading out the wine press' . . . ," reports Montross (p. 138), who observes that these excesses usually came at the climax of the capture of a fortified post or city. "For long the assailants have endured more punishment than they were able to inflict; then once the walls are breached, pent up emotions find an outlet in murder, rape and plunder, which discipline is powerless to prevent." The same occurred when Tyre fell to Alexander after a painful siege, and the phenomenon was not unknown on Pacific islands in the Second World War. Pure violence, like fire, can be harnessed to a purpose; that does not mean that behind every holocaust is a shrewd intention successfully fulfilled.

But if the occurrence of violence does not always bespeak a

shrewd purpose, the absence of pain and destruction is no sign that violence was idle. Violence is most purposive and most successful when it is threatened and not used. Successful threats are those that do not have to be carried out. By European standards, Denmark was virtually unharmed in the Second World War; it was violence that made the Danes submit. Withheld violence—successfully threatened violence—can look clean, even merciful. The fact that a kidnap victim is returned unharmed, against receipt of ample ransom, does not make kidnapping a nonviolent enterprise. The American victory at Mexico City in 1847 was a great success; with a minimum of brutality we traded a capital city for everything we wanted from the war. We did not even have to say what we could do to Mexico City to make the Mexican government understand what they had at stake. (They had undoubtedly got the message a month earlier, when Vera Cruz was being pounded into submission. After forty-eight hours of shellfire, the foreign consuls in that city approached General Scott's headquarters to ask for a truce so that women, children, and neutrals could evacuate the city. General Scott, "counting on such internal pressure to help bring about the city's surrender," refused their request and added that anyone, soldier or noncombatant, who attempted to leave the city would be fired upon.) [2]

Whether spoken or not, the threat is usually there. In earlier eras the etiquette was more permissive. When the Persians wanted to induce some Ionian cities to surrender and join them, without having to fight them, they instructed their ambassadors to

make your proposals to them and promise that, if they abandon their allies, there will be no disagreeable consequences

2. Otis A. Singletary, *The Mexican War* (Chicago, University of Chicago Press, 1960), pp. 75–76. In a similar episode the Gauls, defending the town of Alesia in 52 B.C., "decided to send out of the town those whom age or infirmity incapacitated for fighting. . . . They came up to the Roman fortifications and with tears besought the soldiers to take them as slaves and relieve their hunger. But Caesar posted guards on the ramparts with orders to refuse them admission." Caesar, *The Conquest of Gaul,* S. A. Handford, transl. (Baltimore, Penguin Books, 1951), p. 227.

for them; we will not set fire to their houses or temples, or threaten them with any greater harshness than before this trouble occurred. If, however, they refuse, and insist upon fighting, then you must resort to threats, and say exactly what we will do to them; tell them, that is, that when they are beaten they will be sold as slaves, their boys will be made eunuchs, their girls carried off to Bactria, and their land confiscated.[3]

It sounds like Hitler talking to Schuschnigg. "I only need to give an order, and overnight all the ridiculous scarecrows on the frontier will vanish . . . Then you will really experience something. . . . After the troops will follow the S.A. and the Legion. No one will be able to hinder the vengeance, not even myself."

Or Henry V before the gates of Harfleur:

> We may as bootless spend our vain command
> Upon the enraged soldiers in their spoil
> As send precepts to the leviathan
> To come ashore. Therefore, you men of Harfleur,
> Take pity of your town and of your people,
> Whiles yet my soldiers are in my command;
> Whiles yet the cool and temperate wind of grace
> O'erblows the filthy and contagious clouds
> Of heady murder, spoil and villainy.
> If not, why, in a moment look to see
> The blind and bloody soldier with foul hand
> Defile the locks of your shrill-shrieking daughters;
> Your fathers taken by the silver beard,
> And their most reverent heads dash'd to the walls,
> Your naked infants spitted upon pikes,
> Whiles the mad mothers with their howls confused
> Do break the clouds . . .
> What say you? will you yield, and this avoid,
> Or, guilty in defence, be thus destroy'd?
>
> (Act III, Scene iii)

3. Herodotus, *The Histories,* Aubrey de Selincourt, transl. (Baltimore, Penguin Books, 1954), p. 362.

The Strategic Role of Pain and Damage

Pure violence, nonmilitary violence, appears most conspicuously in relations between unequal countries, where there is no substantial military challenge and the outcome of military engagement is not in question. Hitler could make his threats contemptuously and brutally against Austria; he could make them, if he wished, in a more refined way against Denmark. It is noteworthy that it was Hitler, not his generals, who used this kind of language; proud military establishments do not like to think of themselves as extortionists. Their favorite job is to deliver victory, to dispose of opposing military force and to leave most of the civilian violence to politics and diplomacy. But if there is no room for doubt how a contest in strength will come out, it may be possible to bypass the military stage altogether and to proceed at once to the coercive bargaining.

A typical confrontation of unequal forces occurs at the *end* of a war, between victor and vanquished. Where Austria was vulnerable before a shot was fired, France was vulnerable after its military shield had collapsed in 1940. Surrender negotiations are the place where the threat of civil violence can come to the fore. Surrender negotiations are often so one-sided, or the potential violence so unmistakable, that bargaining succeeds and the violence remains in reserve. But the fact that most of the actual damage was done during the military stage of the war, prior to victory and defeat, does not mean that violence was idle in the aftermath, only that it was latent and the threat of it successful.

Indeed, victory is often but a prerequisite to the exploitation of the power to hurt. When Xenophon was fighting in Asia Minor under Persian leadership, it took military strength to disperse enemy soldiers and occupy their lands; but land was not what the victor wanted, nor was victory for its own sake.

> Next day the Persian leader burned the villages to the ground, not leaving a single house standing, so as to strike terror into the other tribes to show them what would happen if they did

not give in. . . . He sent some of the prisoners into the hills and told them to say that if the inhabitants did not come down and settle in their houses to submit to him, he would burn up their villages too and destroy their crops, and they would die of hunger.[4]

Military victory was but the *price of admission*. The payoff depended upon the successful threat of violence.

Like the Persian leader, the Russians crushed Budapest in 1956 and cowed Poland and other neighboring countries. There was a lag of ten years between military victory and this show of violence, but the principle was the one explained by Xenophon. Military victory is often the prelude to violence, not the end of it, and the fact that successful violence is usually held in reserve should not deceive us about the role it plays.

What about pure violence during war itself, the infliction of pain and suffering as a military technique? Is the threat of pain involved only in the political use of victory, or is it a decisive technique of war itself?

Evidently between unequal powers it has been part of warfare. Colonial conquest has often been a matter of "punitive expeditions" rather than genuine military engagements. If the tribesmen escape into the bush you can burn their villages without them until they assent to receive what, in strikingly modern language, used to be known as the Queen's "protection." British air power was used punitively against Arabian tribesmen in the 1920s and 30s to coerce them into submission.[5]

4. Xenophon, *The Persian Expedition*, Rex Warner, transl. (Baltimore, Penguin Books, 1949), p. 272. "The 'rational' goal of the threat of violence," says H. L. Nieburg, "is an accommodation of interests, not the provocation of actual violence. Similarly the 'rational' goal of actual violence is demonstration of the will and capability of action, establishing a measure of the credibility of future threats, not the exhaustion of that capability in unlimited conflict." "Uses of Violence," *Journal of Conflict Resolution, 7* (1963), 44.

5. A perceptive, thoughtful account of this tactic, and one that emphasizes its "diplomatic" character, is in the lecture of Air Chief Marshal Lord Portal, "Air Force Cooperation in Policing the Empire." "The

If enemy forces are not strong enough to oppose, or are unwilling to engage, there is no need to achieve victory as a prerequisite to getting on with a display of coercive violence. When Caesar was pacifying the tribes of Gaul he sometimes had to fight his way through their armed men in order to subdue them with a display of punitive violence, but sometimes he was virtually unopposed and could proceed straight to the punitive display. To his legions there was more valor in fighting their way to the seat of power; but, as governor of Gaul, Caesar could view enemy troops only as an obstacle to his political control, and that control was usually based on the power to inflict pain, grief, and privation. In fact, he preferred to keep several hundred hostages from the unreliable tribes, so that his threat of violence did not even depend on an expedition into the country-side.

Pure hurting, as a military tactic, appeared in some of the military actions against the plains Indians. In 1868, during the war with the Cheyennes, General Sheridan decided that his best hope was to attack the Indians in their winter camps. His reasoning was that the Indians could maraud as they pleased during the seasons when their ponies could subsist on grass, and in winter hide away in remote places. "To disabuse their minds from the idea that they were secure from punishment, and to strike at a period when they were helpless to move their stock and villages, a winter campaign was projected against the large bands hiding away in the Indian territory." [6]

These were not military engagements; they were punitive attacks on people. They were an effort to subdue by the use of violence, without a futile attempt to draw the enemy's military forces into decisive battle. They were "massive retaliation" on a

---

law-breaking tribe must be given an alternative to being bombed and . . . be told in the clearest possible terms what that alternative is." And, "It would be the greatest mistake to believe that a victory which spares the lives and feelings of the losers need be any less permanent or salutary than one which inflicts heavy losses on the fighting men and results in a 'peace' dictated on a stricken field." *Journal of the Royal United Services Institution* (London, May 1937), pp. 343–58.

6. Paul I. Wellman, *Death on the Prairie* (New York, Macmillan, 1934), p. 82.

diminutive scale, with local effects not unlike those of Hiroshima. The Indians themselves totally lacked organization and discipline, and typically could not afford enough ammunition for target practice and were no military match for the cavalry; their own rudimentary strategy was at best one of harassment and reprisal. Half a century of Indian fighting in the West left us a legacy of cavalry tactics; but it is hard to find a serious treatise on American strategy against the Indians or Indian strategy against the whites. The twentieth is not the first century in which "retaliation" has been part of our strategy, but it is the first in which we have systematically recognized it.

Hurting, as a strategy, showed up in the American Civil War, but as an episode, not as the central strategy. For the most part, the Civil War was a military engagement with each side's military force pitted against the other's. The Confederate forces hoped to lay waste enough Union territory to negotiate their independence, but hadn't enough capacity for such violence to make it work. The Union forces were intent on military victory, and it was mainly General Sherman's march through Georgia that showed a conscious and articulate use of violence. "If the people raise a howl against my barbarity and cruelty, I will answer that war is war . . . If they want peace, they and their relatives must stop the war," Sherman wrote. And one of his associates said, "Sherman is perfectly right . . . The only possible way to end this unhappy and dreadful conflict . . . is to make it terrible beyond endurance." [7]

Making it "terrible beyond endurance" is what we associate with Algeria and Palestine, the crushing of Budapest and the tribal warfare in Central Africa. But in the great wars of the last hundred years it was usually military victory, not the hurting of the people, that was decisive; General Sherman's attempt to make war hell for the Southern people did not come to

7. J. F. C. Fuller reproduces some of this correspondence and remarks, "For the nineteenth century this was a new conception, because it meant that the deciding factor in the war—the power to sue for peace—was transferred from government to people, and that peacemaking was a product of revolution. This was to carry the principle of democracy to its ultimate stage. . . ." *The Conduct of War: 1789–1961* (New Brunswick, Rutgers University Press, 1961), pp. 107–12.

epitomize military strategy for the century to follow. To seek out and to destroy the enemy's military force, to achieve a crushing victory over enemy armies, was still the avowed purpose and the central aim of American strategy in both world wars. Military action was seen as an *alternative* to bargaining, not a *process* of bargaining.

The reason is not that civilized countries are so averse to hurting people that they prefer "purely military" wars. (Nor were all of the participants in these wars entirely civilized.) The reason is apparently that the technology and geography of warfare, at least for a war between anything like equal powers during the century ending in World War II, kept coercive violence from being decisive before military victory was achieved. Blockade indeed was aimed at the whole enemy nation, not concentrated on its military forces; the German civilians who died of influenza in the First World War were victims of violence directed at the whole country. It has never been quite clear whether blockade—of the South in the Civil War or of the Central Powers in both world wars, or submarine warfare against Britain—was expected to make war unendurable for the people or just to weaken the enemy forces by denying economic support. Both arguments were made, but there was no need to be clear about the purpose as long as either purpose was regarded as legitimate and either might be served. "Strategic bombing" of enemy homelands was also occasionally rationalized in terms of the pain and privation it could inflict on people and the civil damage it could do to the nation, as an effort to display either to the population or to the enemy leadership that surrender was better than persistence in view of the damage that could be done. It was also rationalized in more "military" terms, as a way of selectively denying war material to the troops or as a way of generally weakening the economy on which the military effort rested.[8]

8. For a reexamination of strategic-bombing theory before and during World War II, in the light of nuclear-age concepts, see George H. Quester, *Deterrence before Hiroshima* (New York, John Wiley and Sons, 1966). See also the first four chapters of Bernard Brodie, *Strategy in the Missile Age* (Princeton, Princeton University Press, 1959), pp. 3–146.

But as terrorism—as violence intended to coerce the enemy rather than to weaken him militarily—blockade and strategic bombing by themselves were not quite up to the job in either world war in Europe. (They might have been sufficient in the war with Japan after straightforward military action had brought American aircraft into range.) Airplanes could not quite make punitive, coercive violence decisive in Europe, at least on a tolerable time schedule, and preclude the need to defeat or to destroy enemy forces as long as they had nothing but conventional explosives and incendiaries to carry. Hitler's V-1 buzz bomb and his V-2 rocket are fairly pure cases of weapons whose purpose was to intimidate, to hurt Britain itself rather than Allied military forces. What the V-2 needed was a punitive payload worth carrying, and the Germans did not have it. Some of the expectations in the 1920s and the 1930s that another major war would be one of pure civilian violence, of shock and terror from the skies, were not borne out by the available technology. The threat of punitive violence kept occupied countries quiescent; but the wars were won in Europe on the basis of brute strength and skill and not by intimidation, not by the threat of civilian violence but by the application of military force. Military victory was still the price of admission. Latent violence against people was reserved for the politics of surrender and occupation.

The great exception was the two atomic bombs on Japanese cities. These were weapons of terror and shock. They hurt, and promised more hurt, and that was their purpose. The few "small" weapons we had were undoubtedly of some direct military value, but their enormous advantage was in pure violence. In a military sense the United States could gain a little by destruction of two Japanese industrial cities; in a civilian sense, the Japanese could lose much. The bomb that hit Hiroshima was a threat aimed at all of Japan. The political target of the bomb was not the dead of Hiroshima or the factories they worked in, but the survivors in Tokyo. The two bombs were in the tradition of Sheridan against the Comanches and Sherman in Georgia. Whether in the end those two bombs saved lives or

wasted them, Japanese lives or American lives; whether punitive coercive violence is uglier than straightforward military
force or more civilized; whether terror is more or less humane
than military destruction; we can at least perceive that the
bombs on Hiroshima and Nagasaki represented violence against
the country itself and not mainly an attack on Japan's material
strength. The effect of the bombs, and their purpose, were not
mainly the military destruction they accomplished but the pain
and the shock and the promise of more.

The Nuclear Contribution to Terror and Violence

Man has, it is said, for the first time in history enough military
power to eliminate his species from the earth, weapons against
which there is no conceivable defense. War has become, it is
said, so destructive and terrible that it ceases to be an instrument
of national power. "For the first time in human history," says
Max Lerner in a book whose title, *The Age of Overkill,* conveys
the point, "men have bottled up a power . . . which they have thus
far not dared to use." [9] And Soviet military authorities, whose
party dislikes having to accommodate an entire theory of
history to a single technological event, have had to reexamine a
set of principles that had been given the embarrassing name of
"permanently operating factors" in warfare. Indeed, our era is
epitomized by words like "the first time in human history," and
by the abdication of what was "permanent."

For dramatic impact these statements are splendid. Some of
them display a tendency, not at all necessary, to belittle the
catastrophe of earlier wars. They may exaggerate the historical
novelty of deterrence and the balance of terror. [10] More impor-

9.   New York, Simon and Schuster, 1962, p. 47.

10. Winston Churchill is often credited with the term, "balance of terror," and the
following quotation succinctly expresses the familiar notion of nuclear mutual deterrence. This, though, is from a speech in Commons in November 1934. "The fact remains that when all is said and done as regards defensive methods, pending some new
discovery the only direct measure of defense upon a great scale is the certainty of
being able to inflict simultaneously upon the enemy as great damage as he can inflict
upon ourselves. Do not let us undervalue the efficacy of this procedure. It may well

tant, they do not help to identify just what is new about war when so much destructive energy can be packed in warheads at a price that permits advanced countries to have them in large numbers. Nuclear warheads are incomparably more devastating than anything packaged before. What does that imply about war?

It is not true that for the first time in history man has the capability to destroy a large fraction, even the major part, of the human race. Japan was defenseless by August 1945. With a combination of bombing and blockade, eventually invasion, and if necessary the deliberate spread of disease, the United States could probably have exterminated the population of the Japanese islands without nuclear weapons. It would have been a gruesome, expensive, and mortifying campaign; it would have taken time and demanded persistence. But we had the economic and technical capacity to do it; and, together with the Russians or without them, we could have done the same in many populous parts of the world. Against defenseless people there is not much that nuclear weapons can do that cannot be done with an ice pick. And it would not have strained our Gross National Product to do it with ice picks.

It is a grisly thing to talk about. We did not do it and it is not imaginable that we would have done it. We had no reason; if we had had a reason, we would not have the persistence of purpose, once the fury of war had been dissipated in victory and we had taken on the task of executioner. If we and our enemies might do such a thing to each other now, and to others as well,

---

prove in practice—I admit I cannot prove it in theory—capable of giving complete immunity. If two Powers show themselves equally capable of inflicting damage upon each other by some particular process of war, so that neither gains an advantage from its adoption and both suffer the most hideous reciprocal injuries, it is not only possible but it seems probable that neither will employ that means." A fascinating reexamination of concepts like deterrence, preemptive attack, counterforce and countercity warfare, retaliation, reprisal, and limited war, in the strategic literature of the air age from the turn of the century to the close of World War II, is in Quester's book, cited above.

it is not because nuclear weapons have for the first time made it feasible.

Nuclear weapons can do it quickly. That makes a difference. When the Crusaders breached the walls of Jerusalem they sacked the city while the mood was on them. They burned things that they might, with time to reflect, have carried away instead and raped women that, with time to think about it, they might have married instead. To compress a catastrophic war within the span of time that a man can stay awake drastically changes the politics of war, the process of decision, the possibility of central control and restraint, the motivations of people in charge, and the capacity to think and reflect while war is in progress. It *is* imaginable that we might destroy 200,000,000 Russians in a war of the present, though not 80,000,000 Japanese in a war of the past. It is not only imaginable, it is imagined. It is imaginable because it could be done "in a moment, in the twinkling of an eye, at the last trumpet."

This may be why there is so little discussion of how an all-out war might be brought to a close. People do not expect it to be "brought" to a close, but just to come to an end when everything has been spent. It is also why the idea of "limited war" has become so explicit in recent years. Earlier wars, like World Wars I and II or the Franco-Prussian War, were limited by *termination,* by an ending that occurred before the period of greatest potential violence, by negotiation that brought the *threat* of pain and privation to bear but often precluded the massive *exercise* of civilian violence. With nuclear weapons available, the restraint of violence cannot await the outcome of a contest of military strength; restraint, to occur at all, must occur during war itself.

This is a difference between nuclear weapons and bayonets. It is not in the number of people they can eventually kill but in the speed with which it can be done, in the centralization of decision, in the divorce of the war from political processes, and in computerized programs that threaten to take the war out of human hands once it begins.

That nuclear weapons make it *possible* to compress the fury of global war into a few hours does not mean that they make it *inevitable*. We have still to ask whether that is the way a major nuclear war would be fought, or ought to be fought. Nevertheless, that the whole war might go off like one big string of firecrackers makes a critical difference between our conception of nuclear war and the world wars we have experienced.

There is no guarantee, of course, that a slower war would not persist. The First World War could have stopped at any time after the Battle of the Marne. There was plenty of time to think about war aims, to consult the long-range national interest, to reflect on costs and casualties already incurred and the prospect of more to come, and to discuss terms of cessation with the enemy. The gruesome business continued as mechanically as if it had been in the hands of computers (or worse: computers might have been programmed to learn more quickly from experience). One may even suppose it would have been a blessing had all the pain and shock of the four years been compressed within four days. Still, it was terminated. And the victors had no stomach for doing then with bayonets what nuclear weapons could do to the German people today.

There is another difference. In the past it has usually been the victors who could do what they pleased to the enemy. War has often been "total war" for the loser. With deadly monotony the Persians, Greeks, or Romans "put to death all men of military age, and sold the women and children into slavery," leaving the defeated territory nothing but its name until new settlers arrived sometime later. But the defeated could not do the same to their victors. The boys could be castrated and sold only after the war had been won, and only on the side that lost it. The power to hurt could be brought to bear only after military strength had achieved victory. The same sequence characterized the great wars of this century; for reasons of technology and geography, military force has usually had to penetrate, to exhaust, or to collapse opposing military force—to achieve military victory—before it could be brought to bear on the enemy nation itself. The Allies in World War I could not inflict coercive pain and suffering directly on the Germans in a decisive way until they could defeat the German army; and the Germans could not coerce the French people with bayonets unless they first beat the Allied troops that stood in their way. With two-dimensional warfare, there is a tendency for troops to confront each other, shielding their own lands while attempting to press into each other's. Small penetrations could not do major damage to the people; large penetrations were so destructive of military organization that they usually ended the military phase of the war.

Nuclear weapons make it possible to do monstrous violence to the enemy without first achieving victory. With nuclear weapons and today's means of delivery, one expects to penetrate an enemy homeland without first collapsing his military force. What nuclear weapons have done, or appear to do, is to promote this kind of warfare to first place. Nuclear weapons threaten to make war less military, and are responsible for the lowered status of "military victory" at the present time. *Victory is no longer a prerequisite for hurting the enemy.* And it is no assurance against being terribly hurt. One need not wait until he has won the war before inflicting "unendurable" damages on his enemy. One need not wait until he has lost the war. There was a time when the assurance of victory—false or genuine assurance—could make national leaders not just willing but sometimes enthusiastic about war. Not now.

Not only *can* nuclear weapons hurt the enemy before the war has been won, and perhaps hurt decisively enough to make the military engagement academic, but it is widely assumed that in a major war that is *all* they can do. Major war is often discussed as though it would be only a contest in national destruction. If this is indeed the case—if the destruction of cities and their populations has become, with nuclear weapons, the primary object in an all-out war—the sequence of war has been reversed. Instead of destroying enemy forces as a prelude to imposing one's will on the enemy nation, one would have to destroy the nation as a means or a prelude to destroying the enemy forces. If one cannot disable enemy forces without virtually destroying the country, the victor does not even have the option of sparing the conquered nation. He has already destroyed it. Even with blockade and strategic bombing it could be supposed that a country would be defeated before it was destroyed, or would elect surrender before annihilation had gone far. In the Civil War it could be hoped that the South would become too weak to fight before it became too weak to survive. For "all-out" war, nuclear weapons threaten to reverse this sequence.

So nuclear weapons do make a difference, marking an epoch in warfare. The difference is not just in the amount of destruction that can be accomplished but in the role of destruction and in the decision process. Nuclear weapons can change the speed of events, the control of events, the sequence of events, the relation of victor to vanquished, and the relation of homeland to fighting front. Deterrence rests today on the threat of pain and extinction, not just on the threat of military defeat. We may argue about the wisdom of announcing "unconditional surrender" as an aim in the last major war, but seem to expect "unconditional destruction" as a matter of course in another one.

Something like the same destruction always *could* be done. With nuclear weapons there is an expectation that it *would* be done. It is not "overkill" that is new; the American army surely had enough 30 caliber bullets to kill everybody in the world in 1945, or if it did not it could have bought them without any strain. What is new is plain "kill"—the idea that major war might be just a contest in the killing of countries, or not even a contest but just two parallel exercises in devastation.

That is the difference nuclear weapons make. At least they *may* make that difference. They also may not. If the weapons themselves are vulnerable to attack, or the machines that carry them, a successful surprise might eliminate the opponent's means of retribution. That an enormous explosion can be packaged in a single bomb does not by itself guarantee that the victor will receive deadly punishment. Two gunfighters facing each other in a Western town had an unquestioned capacity to kill one another; that did not guarantee that both would die in a gunfight—only the slower of the two. Less deadly weapons, permitting an injured one to shoot back before he died, might have been more conducive to a restraining balance of terror, or of caution. The very efficiency of nuclear weapons could make them ideal for starting war, if they can suddenly eliminate the enemy's capability to shoot back.

And there is a contrary possibility: that nuclear weapons are not vulnerable to attack and prove not to be terribly effective against each other, posing no need to shoot them quickly for fear they will be destroyed before they are launched, and with no task available but the systematic destruction of the enemy country and no necessary reason to do it fast rather than slowly. Imagine that nuclear destruction *had* to go slowly—that the bombs could be dropped only one per day. The prospect would look very different, something like the most terroristic guerila warfare on a massive scale. It happens that nuclear war does not have to go slowly; but it may also not have to go speedily. The mere existence of nuclear weapons does not itself determine that everything must go off in a blinding flash, any more than that it must go slowly. Nuclear weapons do not simplify things quite that much.

In recent years there has been a new emphasis on distinguishing what nuclear weapons make possible and what they make inevitable in case of war. The American government began in 1961 to emphasize that even a major nuclear war might not, and need not, be a simple contest in destructive fury. Secretary McNamara gave a controversial speech in June 1962 on the idea that "deterrence" might operate even in war itself, that belligerents might, out of self-interest, attempt to limit the war's destructiveness. Each might feel the sheer destruction of enemy people and cities would serve no decisive military purpose but that a continued *threat* to destroy them might serve a purpose. The continued threat would depend on their not being destroyed yet. Each might reciprocate the other's restraint, as in limited wars of lesser scope. Even the worst of enemies, in the interest of reciprocity, have often not mutilated prisoners of war; and citizens might deserve comparable treatment. The fury of nuclear attacks might fall mainly on each other's weapons and military forces.

"The United States has come to the conclusion," said Secretary McNamara,

> that to the extent feasible, basic military strategy in a possible general war should be approached in much the same way that more conventional military operations have been regarded in the past. That is to say, principal military objectives . . . should be the destruction of the enemy's military forces, not of his civilian population . . . giving the possible opponent the strongest imaginable incentive to refrain from striking our own cities.[11]

This is a sensible way to think about war, if one has to think about it and of course one does. But whether the Secretary's "new strategy" was sensible or not, whether enemy populations should be held hostage or instantly destroyed, whether the primary targets should be military forces or just people and their source of livelihood, this is not "much the same way that more conventional military operations have been regarded in the past." This is utterly different, and the difference deserves emphasis.

In World Wars I and II one went to work on enemy military forces, not his people, because until the enemy's military forces had been taken care of there was typically not anything decisive that one could do to the enemy nation itself. The Germans did not, in World War I, refrain from bayoneting French citizens by the millions in the hope that the Allies would abstain from shooting up the German population. They could not get at the French citizens until they had breached the Allied lines. Hitler tried to terrorize London and did not make it. The Allied air forces took the war straight to Hitler's territory, with at least some thought of doing in Germany what Sherman recognized he was doing in Georgia; but with the bombing technology of World War II one could not afford to bypass the troops and go exclusively for enemy populations—not, anyway, in Germany. With nuclear weapons one has that alternative.

To concentrate on the enemy's military installations while deliberately holding in reserve a massive capacity for destroying

11. Commencement Address, University of Michigan, June 16, 1962.

his cities, for exterminating his people and eliminating his society, on condition that the enemy observe similar restraint with respect to one's own society, is not the "conventional approach." In World Wars I and II the first order of business was to destroy enemy armed forces because that was the only promising way to make him surrender. To fight a purely military engagement "all-out" while holding in reserve a decisive capacity for violence, on condition the enemy do likewise, is not the way military operations have traditionally been approached. Secretary McNamara was proposing a new approach to warfare in a new era, an era in which the power to hurt is more impressive than the power to oppose.

From Battlefield Warfare to the Diplomacy of Violence

Almost one hundred years before Secretary McNamara's speech, the Declaration of St. Petersburg (the first of the great modern conferences to cope with the evils of warfare) in 1868 asserted, "The only legitimate object which states should endeavor to accomplish during war is to weaken the military forces of the enemy." And in a letter to the League of Nations in 1920, the President of the International Committee of the Red Cross wrote; "The Committee considers it very desirable that war should resume its former character, that is to say, that it should be a struggle between armies and not between populations. The civilian population must, as far as possible, remain outside the struggle and its consequences."[12] His language is remarkably similar to Secretary McNamara's.

The International Committee was fated for disappointment, like everyone who labored in the late nineteenth century to devise rules that would make war more humane. When the Red Cross was founded in 1863, it was concerned about the disregard for noncombatants by those who made war; but in the Second World War noncombatants were deliberately chosen

12. International Committee of the Red Cross, *Draft Rules for the Limitation of the Dangers Incurred by the Civilian Population in Time of War* (2d ed. Geneva, 1958), pp. 144, 151.

as targets by both Axis and Allied forces, not decisively but nevertheless deliberately. The trend has been the reverse of what the International Committee hoped for.

In the present era noncombatants appear to be not only deliberate targets but primary targets, or at least were so taken for granted until about the time of Secretary McNamara's speech. In fact, noncombatants appeared to be primary targets at both ends of the scale of warfare; thermonuclear war threatened to be a contest in the destruction of cities and populations; and, at the other end of the scale, insurgency is almost entirely terroristic. We live in an era of dirty war.

Why is this so? Is war properly a military affair among combatants, and is it a depravity peculiar to the twentieth century that we cannot keep it within decent bounds? Or is war inherently dirty, and was the Red Cross nostalgic for an artificial civilization in which war had become encrusted with etiquette—a situation to be welcomed but not expected?

To answer this question it is useful to distinguish three stages in the involvement of noncombatants—of plain people and their possessions—in the fury of war. These stages are worth distinguishing; but their sequence is merely descriptive of Western Europe during the past three hundred years, not a historical generalization. The first stage is that in which the people may get hurt by inconsiderate combatants. This is the status that people had during the period of "civilized warfare" that the International Committee had in mind.

From about 1648 to the Napoleonic era, war in much of Western Europe was something superimposed on society. It was a contest engaged in by monarchies for stakes that were measured in territories and, occasionally, money or dynastic claims. The troops were mostly mercenaries and the motivation for war was confined to the aristocratic elite. Monarchs fought for bits of territory, but the residents of disputed terrain were more concerned with protecting their crops and their daughters from marauding troops than with whom they owed allegiance to. They were, as Quincy Wright remarked in his classic *Study of War,* little concerned that the territory in which they lived had a new sovereign.[13] Furthermore, as far as the King of Prussia and the Emperor of Austria were concerned, the loyalty and enthusiasm of the Bohemian farmer were not decisive considerations. It is an exaggeration to refer to European war during this period as a sport of kings, but not a gross exaggeration. And the military logistics of those days confined military operations to a scale that did not require the enthusiasm of a multitude.

Hurting people was not a decisive instrument of warfare. Hurting people or destroying property only reduced the value of the things that were being fought over, to the disadvantage of both sides. Furthermore, the monarchs who conducted wars often did not want to discredit the social institutions they shared with their enemies. Bypassing an enemy monarch and taking the war straight to his people would have had revolutionary implications. Destroying the opposing monarchy was often not in the interest of either side; opposing sovereigns had much more in common with each other than with their own subjects, and to discredit the claims of a monarchy might have produced a disastrous backlash. It is not surprising—or, if it is surprising, not altogether astonishing—that on the European continent in that particular era war was fairly well confined to military activity.

One could still, in those days and in that part of the world, be concerned for the rights of noncombatants and hope to devise rules that both sides in the war might observe. The rules might well be observed because both sides had something to gain from preserving social order and not destroying the enemy. Rules might be a nuisance, but if they restricted both sides the disadvantages might cancel out.

This was changed during the Napoleonic wars. In Napoleon's France, people cared about the outcome. The nation was mobilized. The war was a national effort, not just an activity of the elite. It was both political and military genius on the part of Napoleon and his ministers that an entire nation could be mobilized for war. Propaganda became a tool of warfare, and war became vulgarized.

13. Chicago, University of Chicago Press, 1942, p. 296.

Many writers deplored this popularization of war, this involvement of the democratic masses. In fact, the horrors we attribute to thermonuclear war were already foreseen by many commentators, some before the First World War and more after it; but the new "weapon" to which these terrors were ascribed was people, millions of people, passionately engaged in national wars, spending themselves in a quest for total victory and desperate to avoid total defeat. Today we are impressed that a small number of highly trained pilots can carry enough energy to blast and burn tens of millions of people and the buildings they live in; two or three generations ago there was concern that tens of millions of people using bayonets and barbed wire, machine guns and shrapnel, could create the same kind of destruction and disorder.

That was the second stage in the relation of people to war, the second in Europe since the middle of the seventeenth century. In the first stage people had been neutral but their welfare might be disregarded; in the second stage people were involved because it was *their* war. Some fought, some produced materials of war, some produced food, and some took care of children; but they were all part of a war-making nation. When Hitler attacked Poland in 1939, the Poles had reason to care about the outcome. When Churchill said the British would fight on the beaches, he spoke for the British and not for a mercenary army. The war was about something that mattered. If people would rather fight a dirty war than lose a clean one, the war will be between nations and not just between governments. If people have an influence on whether the war is continued or on the terms of a truce, making the war hurt people serves a purpose. It is a dirty purpose, but war itself is often about something dirty. The Poles and the Norwegians, the Russians and the British, had reason to believe that if they lost the war the consequences would be dirty. This is so evident in modern civil wars—civil wars that involve popular feelings—that we expect them to be bloody and violent. To hope that they would be fought cleanly with no violence to people would be a little like hoping for a clean race riot.

There is another way to put it that helps to bring out the sequence of events. If a modern war were a clean one, the violence would not be ruled out but merely saved for the postwar period. Once the army has been defeated in the clean war, the victorious enemy can be as brutally coercive as he wishes. A clean war would determine which side gets to use its power to hurt coercively after victory, and it is likely to be worth some violence to avoid being the loser.

"Surrender" is the process following military hostilities in which the power to hurt is brought to bear. If surrender negotiations are successful and not followed by overt violence, it is because the capacity to inflict pain and damage was successfully used in the bargaining process. On the losing side, prospective pain and damage were averted by concessions; on the winning side, the capacity for inflicting further harm was traded for concessions. The same is true in a successful kidnapping. It only reminds us that the purpose of pure pain and damage is extortion; it is *latent* violence that can be used to advantage. A well-behaved occupied country is not one in which violence plays no part; it may be one in which latent violence is used so skillfully that it need not be spent in punishment.

This brings us to the third stage in the relation of civilian violence to warfare. If the pain and damage can be inflicted during war itself, they need not wait for the surrender negotiation that succeeds a military decision. If one can coerce people and their governments while war is going on, one does not need to wait until he has achieved victory or risk losing that coercive power by spending it all in a losing war. General Sherman's march through Georgia might have made as much sense, possibly more, had the North been losing the war, just as the German buzz bombs and V-2 rockets can be thought of as coercive instruments to get the war stopped before suffering military defeat.

In the present era, since at least the major East–West powers are capable of massive civilian violence during war itself beyond anything available during the Second World War, the occasion for restraint does not await the achievement of military victory or truce. The principal restraint during the Second World War was a temporal boundary, the date of surrender. In the present era we find the violence dramatically restrained during war itself. The Korean War was furiously "all-out" in the fighting, not only on the peninsular battlefield but in the resources used by both sides. It was "all-out," though, only within some dramatic restraints: no nuclear weapons, no Russians, no Chinese territory, no Japanese territory, no bombing of ships at sea or even airfields on the United Nations side of the line. It was a contest in military strength circumscribed by the threat of unprecedented civilian violence. Korea may or may not be a good model for speculation on limited war in the age of nuclear violence, but it was dramatic evidence that the capacity for violence can be consciously restrained even under the provocation of a war that measures its military dead in tens of thousands and that fully preoccupies two of the largest countries in the world.

A consequence of this third stage is that "victory" inadequately expresses what a nation wants from its military forces. Mostly it wants, in these times, the influence that resides in latent force. It wants the bargaining power that comes from its capacity to hurt, not just the direct consequence of successful military action. Even total victory over an enemy provides at best an opportunity for unopposed violence against the enemy population. How to use that opportunity in the national interest, or in some wider interest, can be just as important as the achievement of victory itself; but traditional military science does not tell us how to use that capacity for inflicting pain. And if a nation, victor or potential loser, is going to use its capacity for pure violence to influence the enemy, there may be no need to await the achievement of total victory.

Actually, this third stage can be analyzed into two quite different variants. In one, sheer pain and damage are primary instruments of coercive warfare and may actually be applied, to intimidate or to deter. In the other, pain and destruction *in* war are expected to serve little or no purpose but *prior threats* of sheer violence, even of automatic and uncontrolled violence, are coupled to military force. The difference is in the all-or-none character of deterrence and intimidation. Two acute dilemmas arise. One is the choice of making prospective violence as frightening as possible or hedging with some capacity for reciprocated restraint. The other is the choice of making retaliation as automatic as possible or keeping deliberate control over the fateful decisions. The choices are determined partly by governments, partly by technology. Both variants are characterized by the coercive role of pain and destruction—of threatened (not inflicted) pain and destruction. But in one the threat either succeeds or fails altogether, and any ensuing violence is gratuitous; in the other, progressive pain and damage may actually be used to threaten more. The present era, for countries possessing nuclear weapons, is a complex and uncertain blend of the two.

Coercive diplomacy, based on the power to hurt, was important even in those periods of history when military force was essentially the power to take and to hold, to fend off attack and to expel invaders, and to possess territory against opposition— that is, in the era in which military force tended to pit itself against opposing force. Even then, a critical question was how much cost and pain the other side would incur for the disputed territory. The judgment that the Mexicans would concede Texas, New Mexico, and California once Mexico City was a hostage in our hands was a diplomatic judgment, not a military one. If one could not readily take the particular territory he wanted or hold it against attack, he could take something else and trade it.[14] Judging what the enemy leaders would trade—

14. Children, for example. The Athenian tyrant, Hippias, was besieged in the Acropolis by an army of Athenian exiles aided by Spartans; his position was strong and he had ample supplies of food and drink, and "but for an unexpected accident" says Herodotus, the besiegers would have persevered a while and then retired. But the children of the besieged were caught as they were being taken out of the country for their safety. "This disaster upset all their plans; in order to recover the children, they were forced to accept . . . terms, and agreed to leave Attica within five days." Herodotus, *The Histories,* p. 334. If children can be killed at long distance, by German buzz bombs or nuclear weapons, they do not need to be caught first. And if both can hurt each other's children the bargaining is more complex.

be it a capital city or national survival—was a critical part of strategy even in the past. Now we are in an era in which the power to hurt—to inflict pain and shock and privation on a country itself, not just on its military forces—is commensurate with the power to take and to hold, perhaps more than commensurate, perhaps decisive, and it is even more necessary to think of warfare as a process of violent bargaining. This is not the first era in which live captives have been worth more than dead enemies, and the power to hurt has been a bargaining advantage; but it is the first in American experience when that kind of power has been a dominant part of military relations.

The power to hurt is nothing new in warfare, but for the United States modern technology has drastically enhanced the strategic importance of pure, unconstructive, unacquisitive pain and damage, whether used against us or in our own defense. This in turn enhances the importance of war and threats of war as techniques of influence, not of destruction; of coercion and deterrence, not of conquest and defense; of bargaining and intimidation.

Quincy Wright, in his *Study of War,* devoted a few pages (319–20) to the "nuisance value" of war, using the analogy of a bank robber with a bomb in his hand that would destroy bank and robber. Nuisance value made the threat of war, according to Wright, "an aid to the diplomacy of unscrupulous governments." Now we need a stronger term, and more pages, to do the subject justice, and need to recognize that even scrupulous governments often have little else to rely on militarily. It is extraordinary how many treatises on war and strategy have declined to recognize that the power to hurt has been, throughout history, a fundamental character of military force and fundamental to the diplomacy based on it.

War no longer looks like just a contest of strength. War and the brink of war are more a contest of nerve and risk-taking, of pain and endurance. Small wars embody the threat of a larger war; they are not just military engagements but "crisis diplomacy." The threat of war has always been somewhere underneath international diplomacy, but for Americans it is now much nearer the surface. Like the threat of a strike in industrial relations, the threat of divorce in a family dispute, or the threat of bolting the party at a political convention, the threat of violence continuously circumscribes international politics. Neither strength nor goodwill procures immunity.

Military strategy can no longer be thought of, as it could for some countries in some eras, as the science of military victory. It is now equally, if not more, the art of coercion, of intimidation and deterrence. The instruments of war are more punitive than acquisitive. Military strategy, whether we like it or not, has become the diplomacy of violence.

2

THE ART
OF COMMITMENT

No one seems to doubt that federal troops are available to defend California. I have, however, heard Frenchmen doubt whether American troops can be counted on to defend France, or American missiles to blast Russia in case France is attacked.

It hardly seems necessary to tell the Russians that we should fight them if they attack *us*. But we go to great lengths to tell the Russians that they will have America to contend with if they or their satellites attack countries associated with us. Saying so, unfortunately, does not make it true; and if it is true, saying so does not always make it believed. We evidently do not want war and would only fight if we had to. The problem is to demonstrate that we would have to.

It is a tradition in military planning to attend to an enemy's capabilities, not his intentions. But deterrence is about intentions—not just *estimating* enemy intentions but *influencing* them. The hardest part is communicating our own intentions. War at best is ugly, costly, and dangerous, and at worst disastrous. Nations have been known to bluff; they have also been known to make threats sincerely and change their minds when the chips were down. Many territories are just not worth a war, especially a war that can get out of hand. A persuasive threat of war may deter an aggressor; the problem is to make it persuasive, to keep it from sounding like a bluff.

Military forces are commonly expected to defend their homelands, even to die gloriously in a futile effort at defense. When Churchill said that the British would fight on the beaches nobody supposed that he had sat up all night running once more through the calculations to make sure that that was the right

policy. Declaring war against Germany for the attack on Poland, though, was a different kind of decision, not a simple reflex but a matter of "policy." Some threats are inherently persuasive, some have to be made persuasive, and some are bound to
look like bluffs.

This chapter is about the threats that are hard to make, the
ones that are not inherently so credible that they can be taken
for granted, the ones that commit a country to an action that it
might in somebody's judgment prefer not to take. A good starting point is the national boundary. As a tentative approxima-
tion—a very tentative one—the difference between the
national homeland and everything "abroad" is the difference
between threats that are inherently credible, even if unspoken,
and the threats that have to be made credible. To project the
shadow of one's military force over other countries and territories is an act of diplomacy. To *fight* abroad is a military act,
but to *persuade* enemies or allies that one would fight abroad,
under circumstances of great cost and risk, requires more than a
military capability. It requires projecting intentions. It requires
having those intentions, even deliberately acquiring them, and
communicating them persuasively to make other countries behave.

Credibility and Rationality

It is a paradox of deterrence that in threatening to hurt somebody if he misbehaves, it need not make a critical difference
how much it would hurt you too—*if* you can make him believe
the threat. People walk against traffic lights on busy streets, deterring trucks by walking in front of them.

The principle applied in Hungary in 1956. The West was
deterred by fear of the consequences from entering into what
might have been a legitimate altercation with the Soviet Union
on the proper status of Hungary. The West was deterred not in
the belief that the Soviet Union was stronger than the West or
that a war, if it ensued, would hurt the West more than the Soviet
bloc. The West was deterred because the Soviet Union was
strong enough, and likely enough to react militarily, to make

Hungary seem not worth the risk, no matter who might get hurt worse.

Another paradox of deterrence is that it does not always help to be, or to be believed to be, fully rational, cool-headed, and in control of oneself or of one's country. One of Joseph Conrad's books, *The Secret Agent*, concerns a group of anarchists in London who were trying to destroy bourgeois society. One of their techniques was bomb explosions; Greenwich Observatory was the objective in this book. They got their nitroglycerin from a stunted little chemist. The authorities knew where they got their stuff and who made it for them. But this little purveyor of nitroglycerin walked safely past the London police. A young man who was tied in with the job at Greenwich asked him why the police did not capture him. His answer was that they would not shoot him from a distance—that would be a denial of bourgeois morality, and serve the anarchists' cause—and they dared not capture him physically because he always kept some "stuff" on his person. He kept a hand in his pocket, he said, holding a ball at the end of a tube that reached a container of nitroglycerin in his jacket pocket. All he had to do was to press that little ball and anybody within his immediate neighborhood would be blown to bits with him. His young companion wondered why the police would believe anything so preposterous as that the chemist would actually blow himself up. The little man's explanation was calm. "In the last instance it is character alone that makes for one's safety . . . I have the means to make myself deadly, but that by itself, you understand, is absolutely nothing in the way of protection. What is effective is the belief those people have in my will to use the means. That's their impression. It is absolute. Therefore I am deadly." [1]

We can call him a fanatic, or a faker, or a shrewd diplomatist; but it was worth something to him to have it believed that he would do it, preposterous or not. I have been told that in mental institutions there are inmates who are either very crazy or very wise, or both, who make clear to the attendants that

1. Joseph Conrad, *The Secret Agent* (New York, Doubleday, Page and Company, 1923), pp. 65–68.

they may slit their own veins or light their clothes on fire if they don't have their way. I understand that they sometimes have their way.

Recall the trouble we had persuading Mossadegh in the early 1950s that he might do his country irreparable damage if he did not become more reasonable with respect to his country and the Anglo-Iranian Oil Company. Threats did not get through to him very well. He wore pajamas, and, according to reports, he wept. And when British or American diplomats tried to explain what would happen to his country if he continued to be obstinate, and why the West would not bail him out of his difficulties, it was apparently uncertain whether he even comprehended what was being said to him. It must have been a little like trying to persuade a new puppy that you will beat him to death if he wets on the floor. If he cannot hear you, or cannot understand you, or cannot control himself, the threat cannot work and you very likely will not even make it.

Sometimes we can get a little credit for not having everything quite under control, for being a little impulsive or unreliable. Teaming up with an impulsive ally may do it. There have been serious suggestions that nuclear weapons should be put directly at the disposal of German troops, on the grounds that the Germans would be less reluctant to use them—and that Soviet leaders know they would be less reluctant—than their American colleagues in the early stages of war or ambiguous aggression. And in part, the motive behind the proposals that authority to use nuclear weapons be delegated in peacetime to theater commanders or even lower levels of command, as in the presidential campaign of 1964, is to substitute military boldness for civilian hesitancy in a crisis or at least to make it look that way to the enemy. Sending a high-ranking military officer to Berlin, Quemoy, or Saigon in a crisis carries a suggestion that authority has been delegated to someone beyond the reach of political inhibition and bureaucratic delays, or even of presidential responsibility, someone whose personal reactions will be in a bold military tradition. The intense dissatisfaction of many senators with President Kennedy's restraint over Cuba in early 1962, and with the way matters were left at the close of the crisis in that November, though in many ways an embarrassment to the President, may nevertheless have helped to convey to the Cubans and to the Soviets that, however peaceable the President might want to be, there were political limits to his patience.

A vivid exhibition of national impulsiveness at the highest level of government was described by Averell Harriman in his account of a meeting with Khrushchev in 1959. "Your generals," said Khrushchev, "talk of maintaining your position in Berlin with force. That is bluff." With what Harriman describes as angry emphasis, Khrushchev went on, "If you send in tanks, they will burn and make no mistake about it. If you want war, you can have it, but remember it will be your war. Our rockets will fly automatically." At this point, according to Harriman, Khrushchev's colleagues around the table chorused the word "automatically." The title of Harriman's article in *Life* magazine was, "My Alarming Interview with Khrushchev."[2] The premier's later desk-thumping with a shoe in the hall of the General Assembly was pictorial evidence that high-ranking Russians know how to put on a performance.

General Pierre Gallois, an outstanding French critic of American military policy, has credited Khrushchev with a "shrewd understanding of the politics of deterrence," evidenced by this "irrational outburst" in the presence of Secretary Harriman.[3] Gallois "hardly sees Moscow launching its atomic missiles at Washington because of Berlin" (especially, I suppose, since Khrushchev may not have had any at the time), but apparently thinks nevertheless that the United States ought to appreciate, as Khrushchev did, the need for a kind of irrational automaticity and a commitment to blind and total retaliation.

Even granting, however, that somebody important may be somewhat intimidated by the Russian responsive chorus on automaticity, I doubt whether we want the American government to rely, for the credibility of its deterrent threat, on a corresponding ritual. We ought to get something a little less

2. July 13, 1959, p. 33.
3. *Revue de Défense Nationale,* October 1962.

idiosyncratic for 50 billion dollars a year of defense expenditure. A government that is obliged to appear responsible in its foreign policy can hardly cultivate forever the appearance of impetuosity on the most important decisions in its care. Khrushchev may have needed a short cut to deterrence, but the American government ought to be mature enough and rich enough to arrange a persuasive sequence of threatened responses that are not wholly a matter of guessing a president's temper.

Still, impetuosity, irrationality, and automaticity are not entirely without substance. Displays can be effective, and when President Kennedy took his turn at it people were impressed, possibly even people in the Kremlin. President Kennedy chose a most impressive occasion for his declaration on "automaticity." It was his address of October 22, 1962, launching the Cuban crisis. In an unusually deliberate and solemn statement he said, "Third: it shall be the policy of this nation to regard any nuclear missile launched from Cuba against any nation in the Western hemisphere as an attack by the Soviet Union on the United States, requiring a full retaliatory response upon the Soviet Union." Coming less than six months after Secretary McNamara's official elucidation of the strategy of controlled and flexible response, the reaction implied in the President's statement would have been not only irrational but probably— depending on just what "full retaliatory response" meant to the President or to the Russians—inconsistent with one of the foundations of the President's own military policy, a foundation that was laid as early as his first defense budget message of 1961, which stressed the importance of proportioning the response to the provocation, even in war itself.[4] Nevertheless, it

4. Albert and Roberta Wohlstetter have evaluated this statement of Kennedy's in "Controlling the Risks in Cuba," Adelphi Papers, 17 (London, Institute for Strategic Studies, 1965). They agree that, "This does not sound like a controlled response." They go on to say, "The attempt, it appears, was to say that the United States would respond to a missile against its neighbors as it would respond to one against itself." And this policy, they say, would leave open the possibility of a controlled, or less than "full," reaction. Even if we disregard

was not entirely incredible; and, for all I know, the President meant it.

As a matter of fact it is most unlikely—actually it is inconceivable—that in preparing his address the President sent word to senior military and civilian officials that this particular paragraph of his speech was not to be construed as policy. Even if the paragraph was pure rhetoric, it would probably have been construed in the crisis atmosphere of that eventful Monday as an act of policy. Just affirming such a policy must have made it somewhat more likely that a single atomic explosion in this hemisphere would have been the signal for full-scale nuclear war.

Even if the President had said something quite contrary, had cautioned the Soviets that now was the time for them to take seriously Secretary McNamara's message and the President's own language about proportioning military response to the provocation; if he had served notice that the United States would not be panicked into all-out war by a single atomic event, particularly one that might not have been fully premeditated by the Soviet leadership; his remarks still would not have eliminated the *possibility* that a single Cuban missile, if it contained a nuclear warhead and exploded on the North American continent, could have triggered the full frantic fury of all-out war. While it is hard for a government, particularly a responsible government, to appear irrational whenever such an appearance is expedient, it is equally hard for a government, even a responsible one, to *guarantee* its own moderation in every circumstance.

---

the word "full," though, the threat is still one of nuclear war; and unless we qualify the words, "any nuclear missile," to mean enough to denote deliberate Soviet attack, the statement still has to be classed as akin to Khrushchev's rocket statement, with allowance for differences in style and circumstance. The point is not that the threat was necessarily either a mistake or a bluff, but that it did imply a reaction more readily taken on impulse than after reflection, a "disproportionate" act, one not necessarily serving the national interest if the contingency arose but nevertheless a possibly impressive threat if the government can be credited with that impulse.

All of this may suggest that deterrent threats are a matter of resolve, impetuosity, plain obstinacy, or, as the anarchist put it, sheer character. It is not easy to change our character; and becoming fanatic or impetuous would be a high price to pay for making our threats convincing. We have not the character of fanatics and cannot scare countries the way Hitler could. We have to substitute brains and skill for obstinacy or insanity. (Even then we are at some disadvantage: Hitler had the skill *and* the character—of a sort.)

If we could really make it believed that we would launch general war for every minor infraction of any code of etiquette that we wanted to publish for the Soviet bloc, and if there were high probability that the leaders in the Kremlin knew where their interests lay and would not destroy their own country out of sheer obstinacy, we could threaten anything we wanted to. We could lay down the rules and announce that if they broke any one of them we would inflict the nuclear equivalent of the Wrath of God. The fact that the flood would engulf us, too, is relevant to whether or not the Russians would believe us; but *if* we could make them believe us, the fact that we would suffer too might provide them little consolation.[5] If we could credibly arrange it so that we had to carry out the threat, whether we wished to or not, we would not even be crazy to arrange it so—if we could be sure the Soviets understood the ineluctable consequences of infringing the rules and would have control over themselves. By

5. This is why Gandhi could stop trains by encouraging his followers to lie down on the tracks, and why construction-site integrationists could stop trucks and bulldozers by the same tactic; if a bulldozer can stop more quickly than a prostrate man can get out of its way, the threat becomes fully credible at the point when only the operator of the bulldozer can avert the bloodshed. The same principle is supposed to explain why a less-than-mortal attack on the Soviet Union by a French nuclear force, though exposing France to mortal attack in return, is a deterring prospect to the Soviet Union; credibility is the problem, and some French commentators have proposed legally arranging to put the French force beyond civilian control. American tanks in an anti-riot role may lack credibility, because they threaten too much, as the bulldozer does, even in the use of machine guns to protect each other; so a more credible— a less drastic and fully automatic—device is used to protect the armed steel monsters: a mildly electric bumper.

arranging it so that we might have to blow up the world, we would not have to.

But it is hard to make it believed. It would be hard to keep the Soviets from expecting that we would think it over once more and find a way to give them what my children call "one more chance." Just saying so won't do it. Mossadegh or the anarchist might succeed, but not the American government. What we have to do is to get ourselves into a position where we cannot fail to react as we said we would—where we just cannot help it—or where we would be obliged by some overwhelming cost of not reacting in the manner we had declared.

Coupling Capabilities to Objectives: Relinquishing the Initiative

Often we must maneuver into a position where we no longer have much choice left. This is the old business of burning bridges. If you are faced with an enemy who thinks you would turn and run if he kept advancing, and if the bridge is there to run across, he may keep advancing. He may advance to the point where, if you do not run, a clash is automatic. Calculating what is in your long-run interest, you may turn and cross the bridge. At least, he may expect you to. But if you burn the bridge so that you cannot retreat, and in sheer desperation there is nothing you can do but defend yourself, he has a new calculation to make. He cannot count on what you would *prefer* to do if he were advancing irresistibly; he must decide instead what he ought to do if you were incapable of anything but resisting him.

This is the position that Chiang Kai-shek got himself into, and us with him, when he moved a large portion of his best troops to Quemoy. Evacuation under fire would be exceedingly difficult; if attacked, his troops had no choice but to fight, and we probably had no choice but to assist them. It was undoubtedly a shrewd move from Chiang's point of view—coupling himself, and the United States with him, to Quemoy—and in fact if we had wanted to make clear to the Chinese Communists that Quemoy had to be defended if they attacked it, it would even have been a shrewd move also from our point of view.

This idea of burning bridges—of maneuvering into a position where one clearly cannot yield—conflicts somewhat, at least semantically, with the notion that what we want in our foreign policy is "the initiative." Initiative is good if it means imaginativeness, boldness, new ideas. But the term somewhat disguises the fact that deterrence, particularly deterrence of anything less than mortal assault on the United States, often depends on getting into a position where the initiative is up to the enemy and it is he who has to make the awful decision to proceed to a clash.

In recent years it has become something of a principle in the Department of Defense that the country should have abundant "options" in its choice of response to enemy moves. The principle is a good one, but so is a contrary principle—that certain options are an embarrassment. The United States government goes to great lengths to reassure allies and to warn Russians that it has eschewed certain options altogether, or to demonstrate that it could not afford them or has placed them out of reach. The *commitment* process on which all American overseas deterrence depends—and on which all confidence within the alliance depends—is a process of surrendering and destroying options that we might have been expected to find too attractive in an emergency. We not only give them up in exchange for commitments *to* us by our allies; we give them up on our own account to make our intentions clear to potential enemies. In fact, we do it not just to display our intentions but to *adopt* those intentions. If deterrence fails it is usually because someone thought he saw an "option" that the American government had failed to dispose of, a loophole that it hadn't closed against itself.

At law there is a doctrine of the "last clear chance." It recognizes that, in the events leading up to an accident, there was some point prior to which either party could avert collision, some point after which neither could, and very likely a period between when one party could still control events but the other was helpless to turn aside or stop. The one that had the "last clear chance" to avert collision is held responsible. In strategy when both parties abhor collision the advantage goes often to the one who arranges the status quo in his favor and leaves to the other the "last clear chance" to stop or turn aside. Xenophon understood the principle when, threatened by an attack he had not sought, he placed his Greeks with their backs against an impassable ravine. "I should like the enemy to think it is easy-going in every direction for *him* to retreat." And when he had to charge a hill occupied by aliens, he "did not attack from every direction but left the enemy a way of escape, if he wanted to run away." The "last chance" to clear out was left to the enemy when Xenophon had to take the initiative, but denied to himself when he wanted to deter attack, leaving his enemy the choice to attack or retire.[6]

An illustration of this principle—that deterrence often depends on relinquishing the initiative to the other side—may be found in a comparison of two articles that Secretary Dulles wrote in the 1950s. His article in *Foreign Affairs* in 1954 (based on the speech in which he introduced "massive retaliation") proposed that we should not let the enemy know in advance just when and where and how we would react to aggression, but reserve for *ourselves* the decision on whether to act and the time, place, and scope of our action. In 1957 the Secretary wrote another article in *Foreign Affairs,* this one oriented mainly toward Europe, in which he properly chose to reserve for the Soviets the final decision on all-out war. He discussed the need for more powerful NATO forces, especially "tactical" nuclear forces that could resist a non-nuclear Soviet onslaught at a level short of all-out war. He said:

> In the future it may thus be feasible to place less reliance upon deterrence of vast retaliatory power. . . . Thus, in

6. *The Persian Expedition,* pp. 136–37, 236. The principle was expressed by Sun Tzu in China, around 500 B.C. in his *Art of War:* "When you surround an army leave an outlet free. Do not press a desperate foe too hard." Ptolemy, serving under Alexander in the fourth century B.C., surrounded a hill, "leaving a gap in his line for the enemy to get through, should they wish to make their escape." Vegetius, writing in the fourth century A.D., had a section headed, "The flight of an enemy should not be prevented, but facilitated," and commends a maxim of Scipio "that a golden bridge should be made for a flying enemy." It is, of course, a fundamental principle of riot control and has its counterparts in diplomacy and other negotiations.

contrast to the 1950 decade, it may be that by the 1960 decade the nations which are around the Sino-Soviet perimeter can possess an effective defense against full-scale conventional attack and thus confront any aggressor with the choice between failing or himself initiating nuclear war against the defending country. Thus the tables may be turned, in the sense that instead of those who are non-aggressive having to rely upon all-out nuclear retaliatory power for their protection, would-be aggressors will be unable to count on a successful conventional aggression, but must themselves weigh the consequences of invoking nuclear war.[7]

Former Secretary Dean Acheson was proposing the same principle (but attached to conventional forces, not tactical nuclear weapons) in remarkably similar language at about the same time in his book, *Power and Diplomacy:*

> Suppose, now, that a major attack is mounted against a Western Europe defended by substantial and spirited forces including American troops. . . . Here, in effect, he (our potential enemy) would be making the decision for us, by compelling evidence that he had determined to run all risks and force matters to a final showdown, including (if it had not already occurred) a nuclear attack upon us. . . . A defense in Europe of this magnitude will pass the decision to risk everything from the defense to the offense.[8]

The same principle on the Eastern side was reflected in a remark often attributed to Khrushchev. It was typically agreed, especially at summit meetings, that nobody wanted a war. Khrushchev's complacent remark, based on Berlin's being on his side of the border, was that Berlin was not worth a war. As the story goes, he was reminded that Berlin was not worth a war to him either. "No," he replied, "but you are the ones that have

7. "Challenge and Response in U.S. Foreign Policy," *Foreign Affairs, 36* (1957), 25–43. It is interesting that Secretary Dulles used "nuclear war" to mean something that had not yet been invoked when "tactical" nuclear weapons were already being used in local defense of Europe.

8. Cambridge, Harvard University Press, 1958, pp. 87–88.

to cross a frontier." The implication, I take it, was that neither of us wanted to cross that threshold just for Berlin, and if Berlin's location makes *us* the ones who have to cross the border, we are the ones who let it go though both of us are similarly fearful of war.

How do we maneuver into a position so it is the other side that has to make that decision? Words rarely do it. To have told the Soviets in the late 1940s that, if they attacked, we were obliged to defend Europe might not have been wholly convincing. When the Administration asked Congress for authority to station Army divisions in Europe in peacetime, the argument was explicitly made that these troops were there not to defend against a superior Soviet army but to leave the Soviet Union in no doubt that the United States would be automatically involved in the event of any attack on Europe. The implicit argument was not that since we obviously would defend Europe we should demonstrate the fact by putting troops there. The reasoning was probably that, whether we wished to be or not, we could not fail to be involved if we had more troops being run over by the Soviet Army than we could afford to see defeated. Notions like "trip wire" or "plate glass window," though oversimplified, were attempts to express this role. And while "trip wire" is a belittling term to describe an army, the role is not a demeaning one. The garrison in Berlin is as fine a collection of soldiers as has ever been assembled, but excruciatingly small. What can 7,000 American troops do, or 12,000 Allied troops? Bluntly, they can die. They can die heroically, dramatically, and in a manner that guarantees that the action cannot stop there. They represent the pride, the honor, and the reputation of the United States government and its armed forces; and they can apparently hold the entire Red Army at bay. Precisely because there is no graceful way out if we wished our troops to yield ground, and because West Berlin is too small an area in which to ignore small encroachments, West Berlin and its military forces constitute one of the most impregnable military outposts of modern times. The Soviets have not dared to cross *that* frontier.

Berlin illustrates two common characteristics of these commitments. The first is that if the commitment is ill defined and ambiguous—if we leave ourselves loopholes through which to exit—our opponent will expect us to be under strong temptation to make a graceful exit (or even a somewhat graceless one) and he may be right. The western sector of Berlin is a tightly defined piece of earth, physically occupied by Western troops: our commitment is credible because it is inescapable. (The little enclave of Steinstücken is physically separate, surrounded by East German territory outside city limits, and there has been a certain amount of jockeying to determine how credible our commitment is to stay there and whether it applies to a corridor connecting the enclave to the city proper.) But our commitment to the integrity of Berlin itself, the entire city, was apparently weak or ambiguous. When the Wall went up the West *was* able to construe its obligation as *not* obliging forceful opposition. The Soviets probably anticipated that, if the West had a choice between interpreting its obligation to demand forceful opposition and interpreting the obligation more leniently, there would be a temptation to elect the lenient interpretation. If we could have *made* ourselves obliged to knock down the wall with military force, the wall might not have gone up; not being obliged, we could be expected to elect the less dangerous course.

The second thing that Berlin illustrates is that, however precisely defined is the issue *about* which we are committed, it is often uncertain just what we are committed *to do*. The commitment is open-ended. Our military reaction to an assault on West Berlin is really not specified. We are apparently committed to holding the western sector of the city if we can; if we are pushed back, we are presumably committed to repelling the intruders and restoring the original boundary. If we lose the city, we are perhaps committed to reconquering it. But somewhere in this sequence of events things get out of hand, and the matter ceases to be purely one of restoring the status quo in Berlin. Military instabilities may arise that make the earlier status quo meaningless. A costly reestablishment of the status quo might call for some sort of reprisal, obliging some counteraction in return. Just what would happen is a matter of prediction, or guess.

What we seem to be committed to is action of some sort commensurate with the provocation. Military resistance tends to develop a momentum of its own. It is dynamic and uncertain. What we threaten in Berlin is to initiate a process that may quickly get out of hand.

The maneuver in Lebanon in 1958—the landing of troops in a developing crisis—though not one of the neatest political–military operations of recent times, represented a similar strategy. Whatever the military potential of the ten or twelve thousand troops that we landed in Lebanon—and it would depend on who might have engaged them, where, over what issue—they had the advantage that they got on the ground before any Soviet adventure or movement was under way. The landing might be described as a "preemptive maneuver." From then on, any significant Soviet intervention in the affairs of Lebanon, Jordan, or even Iraq, would have substantially raised the likelihood that American and Soviet forces, or American and Soviet-supported forces, would be directly engaged.

In effect, it was Khrushchev's turn to cross a border. Iraq or Jordan might not have been worth a war to either of us but by getting troops on the soil—or, as we used to say, the American flag—we probably made it clear to the Kremlin that we could not gracefully retreat under duress. It is harder to retreat than not to land in the first place; the landing helped to put the next step up to the Russians.

Coupling Capabilities to Objectives:
The Process of "Commitment"

In addition to getting yourself where you *cannot* retreat, there is a more common way of making a threat. That is to incur a political involvement, to get a nation's honor, obligation, and diplomatic reputation committed to a response. The Formosa resolution of 1955, along with the military assistance agreement then signed by the United States and the National Government of the Republic of China, should probably be interpreted that way. It was not mainly a technique for reassuring Chiang Kai-shek that we would defend him, and it was not mainly a quid pro quo for something he did for us. It was chiefly important as a move to impress a third party. The primary audience for the congressional action was inside the Soviet bloc. The resolution, together with the treaty, was a ceremony to leave the Chinese and the Russians under no doubt that we could not back down from the defense of Formosa without intolerable loss of prestige, reputation, and leadership. We were not merely *communicating* an intention or obligation we already had, but actually *enhancing* the obligation in the process. The congressional message was not, "Since we are obliged to defend Formosa, we may as well show it." Rather: "In case we were not sufficiently committed to impress you, now we are. We hereby oblige ourselves. Behold us in the public ritual of getting ourselves genuinely committed." [9]

9. There is also sometimes available an internal technique of commitment. It is, in the words of Roger Fisher, "to weave international obligations into the domestic law of each country, so that by and large each government enforces the obligation against itself." Fisher discussed it in relation to disarmament commitments; but it may apply to the use of force as well as to the renunciation of it. A Norwegian directive *(Kgl res 10 Juni 1949)* stipulates that, in event of armed attack, military officers are to mobilize whether or not the government issues the order, that orders for discontinuance issued in the name of the government shall be assumed false, and that resistance is to continue irrespective of enemy threats of retaliatory bombing. Similarly a Swiss order of April 1940, distributed to every soldier in his *livret de service,* declared that in event of attack the Swiss would fight and that any order or indication to the contrary, from any source, was to be considered enemy propaganda. The purposes appear to have been internal discipline and morale; but the possible contribution of such internal arrangements to deterrence, to the credibility of resistance, is worth considering. Many governments have had constitutional or informal provisions for increasing the authority of the armed forces in time of emergency, thus possibly shifting government authority in the direction of individuals and organizations whose motives to resist were less doubtful. As mentioned in an earlier footnote, legal automaticity has sometimes been proposed for the French nuclear force. Internal public opinion can be similarly manipulated to make accommodation unpopular. All of these techniques, if appreciated by the enemy to be deterred, are relevant to the process of commitment. They can also, of course, be quite dangerous. Fisher's discussion is in his chapter, "Internal Enforcement of International Rules," *Disarmament: Its Politics and Economics,* Seymour Melman, ed. (Boston, American Academy of Arts and Sciences, 1962).

That kind of commitment is not to be had cheaply. If Congress passed such a resolution for every small piece of the world that it would like the Soviets to leave alone, it would cheapen the currency. A nation has limited resources, so to speak, in the things that it can get exceptionally concerned about. Political involvement within a country is not something that can be had for the price of a casual vote or a signature on a piece of paper.

Sometimes it comes about by a long process that may not even have been deliberately conceived. As far as I can tell, we had only the slightest commitment, if any, to assist India in case of attack by the Chinese or the Russians, if only because over the years the Indians did not let us incur a formal commitment. One of the lessons of November 1962 may be that, in the face of anything quite as adventuresome as an effort to take over a country the size of India, we may be virtually as committed as if we had a mutual assistance treaty. We cannot afford to let the Soviets or Communist Chinese learn by experience that they can grab large chunks of the earth and its population without a genuine risk of violent Western reaction.

Our commitment to Quemoy, which gave us concern in 1955 and especially in 1958, had not been deliberately conceived; and it appeared at the time to be a genuine embarrassment. For reasons that had nothing to do with American policy, Quemoy had been successfully defended by the Nationalists when Chiang Kai-shek evacuated the mainland, and it remained in Nationalist hands. By the time the United States assumed the Commitment to Formosa, the island of Quemoy stood as a ragged edge about which our intentions were ambiguous. Secretary Dulles in 1958 expressed the official view that we could not afford to vacate Quemoy under duress. The implication seemed to be that we had no genuine desire to take risks for Quemoy and might have preferred it if Quemoy had fallen to the Communists in 1949; but our relations with Communist China were at stake once Quemoy became an issue. So we had a commitment that we might have preferred not to have. And in case that commitment did not appear firm enough, Chiang Kai-shek increased it for us by moving enough of his best troops to that island, under conditions in which evacuation under attack would have been difficult, to make clear that *he* had to defend it or suffer military disaster, leaving it up to the United States to bail him out.

Some of our strongest commitments may be quite implicit, though ritual and diplomacy can enhance or erode them. Commitments can even exist when we deny them. There is a lot of conjecture about what would happen if the NATO treaty lapsed after its initial twenty years. There has recently been some conjecture whether the developing community of Western Europe might be inconsistent with the Atlantic Alliance. It is sometimes argued that the Soviet Union would like Europe so self-reliant that the United States could ease itself out of its commitments to the present NATO countries. I think there is something in this—our commitment to Europe probably diminishes somewhat if the NATO treaty legally goes out of force—but not much. Most of the commitment will still be there. We cannot afford to let the Soviets overrun West Germany or Greece, irrespective of our treaty commitments to Germany or to the rest of Western Europe.

I suspect that we might even recognize an implicit obligation to support Yugoslavia, perhaps Finland, in a military crisis. Any commitment we may have had toward Hungary was apparently not much. But Yugoslavia and Finland have not quite the status that Hungary had. (Conceivably we might cross the border first, under invitation, and leave it up to the Soviets to decide whether to incur the risk of engaging us.) I wonder whether the Kremlin thinks that, if it should get genuinely impatient with Tito or if there were some kind of crisis of succession upon Tito's death, the Red Army could simply invade Yugoslavia or the Kremlin present an ultimatum to the country without any danger of a counter-ultimatum from us or another preemptive landing of troops as in Lebanon. I can only wonder; these are all matters of interpretation, both as to what our commitments really would prove to be and what the Soviets would believe them to be.

Actually, our commitment is not so much a policy as a prediction. We cannot have a clear policy for every contingency; there are too many contingencies and not enough hours in the day to work them all out in advance. If one had asked in October 1962 what American policy was for the contingency of a Communist Chinese effort to destroy the Indian Army, the only answer could have been a *prediction* of what the American government *would* decide to do in a contingency that probably had not been "staffed out" in advance. Policy is usually not a prefabricated decision; it is the whole set of motives and constraints that make a government's actions somewhat predictable.

In the Indian case, it turns out that we had a latent or implicit policy. For all I know, Mr. Nehru anticipated it for ten years. It is conceivable—though I doubt it—that one of the reasons Nehru was so contemptuous of the kinds of treaties that the Thai and Pakistani signed with us was that he felt that his own involvement with the West in a real emergency might be about as strong without the treaty as with it. It is interesting that any "commitment" we had to keep India from being conquered or destroyed by Communist China was not mainly a commitment *to* the Indians or their government. We wanted to restrain Communist China generally; we wanted to give confidence to other governments in Asia; and we wanted to preserve confidence in our deterrent role all the way around the world to Europe. Military support to India would be a way of keeping an implicit pledge but the pledge was a general one, not a debt owed to the Indians. When a disciplinarian—police or other—intervenes to resist or punish someone's forbidden intrusion or assault, any benefit to the victim of the intrusion or assault may be incidental. He could even prefer not to be fought over; but if the issue is maintenance of discipline, he may not have much say in the matter.

This matter of prediction may have been crucial at the start of the Korean War. There has been a lot of discussion about whether we were or were not "committed" to the defense of South Korea. From what I have seen of the way the decision to intervene was taken, first by participation of American military assistance forces, then by bombing, then with reinforcements, and finally with a major war effort, one could not confidently have guessed in May 1950 what the United States would do. One could only try to estimate the probable decision that the President would take, depending on what it looked like in Korea, who was advising him, and what else was going on in the world.

You will recall discussion about the importance of a particular speech by Secretary of State Acheson in suggesting to the Soviets that South Korea was outside our defense perimeter. (As far as I know, there is no decisive evidence that the Russians, Chinese, or Koreans were particularly motivated by that statement.) His stated position was essentially that we had a defense perimeter that excluded South Korea, that we had various other obligations, especially to the United Nations, that would cover a country like South Korea. Apparently the Soviets (or Chinese, or whoever made the decision) miscalculated; they may have thought we were damning our commitment with faint praise. They got into an expensive war and a risky one and one that might have been even more dangerous than it was. They may have miscalculated because the language of deterrence, and an understanding of the commitment process in the nuclear era, had not had much time to develop yet. They may interpret better now—although the missile adventure in Cuba shows that the Soviets could still misread the signals (or the Americans could still fail to transmit them clearly) a decade later.

And we seem to have misread the Chinese warnings during our advance toward the Yalu River. Allen Whiting has documented a serious Chinese Communist attempt to warn the Americans that they would engage us militarily rather than let us occupy all of North Korea.[10] Whatever we might have done had we understood them, we manifestly did not understand. The one thing we would not have done, had we received their warnings correctly, was to extend our forces as vulnerably as we did. We either did not get their message, did not comprehend it, or

10. *China Crosses the Yalu* (New York, Macmillan, 1960).

did not find it credible, though the Chinese Communists may have been doing the best they could to get the message to us and to make it credible. When communication fails, it is not easy to decide whether the transmitter is too weak for the receiver or the receiver too weak for the transmitter, whether the sender speaks the receiver's language badly or the receiver misunderstands the sender's. Between the two of us, Americans and Communist China, we appear to have suffered at least one communication failure in each direction in 1950.[11]

The Interdependence of Commitments

The main reason why we are committed in many of these places is that our threats are interdependent. Essentially we tell the Soviets that we have to react here because, if we did not, they would not believe us when we say that we will react there.

By now our commitment to Berlin has become so deep and diffuse that most of us do not often have to think about whom our commitment is to. The reason we got committed to the defense of Berlin, and stayed committed, is that if we let the Soviets scare us out of Berlin we would lose face with the Soviets themselves. The reputation that most matters to us is our reputation with the Soviet (and Communist Chinese) leaders. It would be bad enough to have Europeans, Latin Americans, or Asians think that we are immoral or cowardly. It would be far worse to lose our reputation with the Soviets. When we talk about the loss of face that would occur if we backed out of For-

11. It is not easy to explain why the Chinese entered North Korea so secretly and so suddenly. Had they wanted to stop the United Nations forces at the level, say, of Pyongyang, to protect their own border and territory, a conspicuous early entry in force might have found the U.N. Command content with its accomplishment and in no mood to fight a second war, against Chinese armies, for the remainder of North Korea. They chose instead to launch a surprise attack, with stunning tactical advantages but no prospect of deterrence. It may have been a hard choice with the decision, finally, a pessimistic one; if so, it was probably a mistake. It may have been based on an overriding interest in the territorial integrity of a Communist North Korea; if so, accommodation was probably impossible anyhow. Or it may have been just a military obsession with tactical surprise, at the expense of all deterrence and diplomacy.

mosa under duress, or out of Berlin, the loss of face that matters most is the loss of Soviet belief that we will do, elsewhere and subsequently, what we insist we will do here and now. Our deterrence rests on Soviet expectations.

This, I suppose, is the ultimate reason why we have to defend California—aside from whether or not Easterners want to. There is no way to let California go to the Soviets and make them believe nevertheless that Oregon and Washington, Florida and Maine, and eventually Chevy Chase and Cambridge cannot be had under the same principle. There is no way to persuade them that if we do not stop them in California we will stop them at the Mississippi (though the Mississippi is a degree less implausible than any other line between that river and, say, the continental divide). Once they cross a line into a new class of aggression, into a set of areas or assets that we always claimed we would protect, we may even deceive *them* if we do not react vigorously. Suppose we let the Soviets have California, and when they reach for Texas we attack them in full force. They could sue for breach of promise. We virtually told them they could have Texas when we let them into California; the fault is ours, for communicating badly, for not recognizing what we were conceding.

California is a bit of fantasy here; but it helps to remind us that the effectiveness of deterrence often depends on attaching to particular areas some of the status of California. The principle is at work all over the world; and the principle is not wholly under our own control. I doubt whether we can identify ourselves with Pakistan in quite the way we can identify ourselves with Great Britain, no matter how many treaties we sign during the next ten years.

"To identify" is a complex process. It means getting the Soviets or the Communist Chinese to identify us with, say, Pakistan in such a way that *they* would lose respect for our commitments elsewhere if we failed to support Pakistan and *we* know they would lose that respect, so that we would have to support Pakistan and *they* know we would. In a way, it is the Soviets who confer this identification; but they do it through the medium of their expectations about us and our understanding of their expectations. Neither they nor we can exercise full control over their expectations.

There is an interesting geographical difference in the Soviet and American homelands; it is hard to imagine a war so located that it could spill over by hot pursuit, by interdiction bombing, by inadvertent border violation, by local reprisal bombing, or even by deliberate but limited ground encroachment into American territory. Our oceans may not protect us from big wars but they protect us from little ones. A local war could not impinge on California, involving it peripherally or incidentally through geographical continuity, the way the Korean War could impinge on Manchuria and Siberia, or the way Soviet territory could be impinged on by war in Iran, Yugoslavia, or Central Europe. One can argue about how far back toward Moscow an "interdiction campaign" of bombing might have to reach, or might safely reach, in case of a limited war in Central Europe; and there is no geographical feature—and few economic features —to present a sudden discontinuity at the Soviet border. A comparable question hardly arises for American participation in the same war; there is one discontinuity leading to submarine warfare on the high seas, and another, a great one, in going inland to the railroad tracks that carry the freight to the Baltimore docks. The vehicles or vessels that would have to carry out the intrusion would furthermore be different in character from those involved in the "theater war."

Possibilities of limited, marginal, homeland engagement that might be logically pertinent for California or Massachusetts are just geographically inapplicable. This gives the American homeland a more distinctive character—a more unambiguous "homeland" separateness—than the Soviet homeland can have. The nearest thing to "local involvement" one can imagine might be Florida bases in case of an air war with Cuba; that would be a possible exception to the rule, while for the Soviet Union most of the hypothetical wars that they must have to make plans about raise the problem of peripheral homeland involvement of some sort (including intrusive reconnaissance and other airspace violations even if no dirt is disturbed on their territory).

The California principle actually can apply not only to terri-tories but to weapons. One of the arguments that has been made, and taken seriously, against having all of our strategic weapons at sea or in outer space or even emplaced abroad, is that the enemy might be able to attack them without fearing the kind of response that would be triggered by an attack on our homeland. If all missiles were on ships at sea, the argument runs, an attack on a ship would not be quite the same as an at-tack on California or Massachusetts; and an enemy might consider doing it in circumstances when he would not consider attacking weapons located on our soil. (An extreme form of the argument, not put forward quite so seriously, was that we ought to locate our weapons in the middle of population centers, so that the enemy could never attack them without arousing the massive response that he could take for granted if he struck our cities.)

There is something to the argument. If in an Asian war we flew bombers from aircraft carriers or from bases in an allied country, and an enemy attacked our ships at sea or our overseas bases, we would almost certainly not consider it the same as if we had flown the bombers from bases in Hawaii or California and he had attacked the bases in those states. If the Soviets had put nuclear weapons in orbit and we shot at them with rockets the results might be serious, but not the same as if the Soviets had put missiles on home territory and we shot at those missiles on their home grounds. Missiles in Cuba, though owned and manned by Russians, were less "nationalized" as a target than missiles in the U.S.S.R. itself. (One of the arguments made against the use of surface ships in a European Multilateral Force armed with long-range missiles was that they could be picked off by an enemy, possibly during a limited war in which the Multilateral Force was not engaged, possibly without the use of nuclear weapons by an enemy, in a way that would not quite provoke reprisal, and thus would be vulnerable in a way that homeland-based missiles would not be.)

The argument can go either way. This can be a reason for deliberately putting weapons outside our boundary, so that their military involvement would not tempt an attack on our homeland, or for keeping them within our boundaries so that an attack on them would appear the more risky. The point here is just that there is a difference. Quemoy cannot be made part of California by moving it there, but weapons can.

Actually the all-or-nothing character of the homeland is not so complete. Secretary McNamara's suggestion that even a general war might be somewhat confined to military installations, and that a furious attack on enemy population centers might be the proper response only to an attack on ours, implies that we do distinguish or might distinguish different parts of our territory by the degree of warfare involved. And I have heard it argued that the Soviets, if they fear for the deterrent security of their retaliatory forces in a purely "military" war that the Americans might initiate, may actually prefer a close proximity of their missiles to their cities to make the prospect of a "clean" strategic war, one without massive attacks on cities, less promising—to demonstrate that there would remain little to lose, after an attack on their weapons, and little motive to confine their response to military targets. The policy would be a dangerous one if there were much likelihood that war would occur, but its logic has merit.

Discrediting an Adversary's Commitments

The Soviets have the same deterrence problem beyond their borders that we have. In some ways the West has helped them to solve it. All kinds of people, responsible and irresponsible, intelligent and unenlightened, European and American, have raised questions about whether the United States really would use its full military force to protect Western Europe or to retaliate for the loss of Western Europe. Much more rarely did I hear anyone question—at least before about 1963—whether the Soviets would do likewise if we were provoked to an attack against the homeland of Communist China.

The Soviets seem to have accomplished—and we helped

them—what we find difficult, namely, to persuade the world that the entire area of their alliance is part of an integral bloc. In the West we talked for a decade—until the Sino-Soviet schism became undeniable—about the Sino-Soviet bloc as though every satellite were part of the Soviet system, and as though Soviet determination to keep those areas under their control was so intense that they could not afford to lose any of it. We often acted as though every part of their sphere of influence was a "California." In the West we seemed to concede to the Soviets, with respect to China, what not everybody concedes us with respect to Europe.

If we always treat China as though it is a Soviet California, we tend to make it so. If we imply to the Soviets that we consider Communist China or Czechoslovakia the virtual equivalent of Siberia, then in the event of any military action in or against those areas we have informed the Soviets that we are going to interpret their response as though we had landed troops in Vladivostok or Archangel or launched them across the Soviet-Polish border. We thus *oblige* them to react in China, or in North Vietnam or wherever it may be, and in effect give them precisely the commitment that is worth so much to them in deterring the West. If we make it clear that we believe they are obliged to react to an intrusion in Hungary as though we were in the streets of Moscow, then they *are* obliged.

Cuba will continue to be an interesting borderline case. The Soviets will find it difficult politically and psychologically to get universal acquiescence that a country can be genuinely within the Soviet bloc if it is not contiguous to them. The Soviet problem was to try to get Cuba into the status of a Soviet "California." It is interesting to speculate on whether we could add states to the Union, like the Philippines, Greece, or Formosa, and let that settle the question of where they belong and how obliged we are to defend them. Hawaii, yes, and by now Puerto Rico; but if we reached out beyond the areas that "belong" in the United States we could probably just not manage to confer a genuinely plausible "statehood" that would be universally recognized and taken for granted.

And Cuba does not quite "belong" in the Soviet bloc—it is topologically separate and does not enjoy the territorial integrity with the Soviet bloc that nations traditionally enjoy. India could take Goa for what are basically esthetic reasons: a conventional belief that maps ought to have certain geometrical qualities, that an enclave is geographically abnormal, that an island in the ocean can belong to anyone but an island surrounded by the territory of a large nation somehow ought to belong to it. (Algeria would, for the same reason, have been harder to disengage from metropolitan France had it not been geographically separated by the Mediterranean; keeping the coastal cities in "France" while dividing off the hinterland would similarly have gone somewhat against cartographic psychology.) There are many other things, of course, that make Cuba different from Hungary, including the fact that the United States can surround it, harass it, or blockade it without encroaching on Soviet territory. But even without that it would be an uphill struggle for the Soviets to achieve a credible togetherness with the remote island of Cuba.

Additional "Cubas" would cost the Soviets something. That does not mean we should like them; still, we should recognize what happens to their deterrence problem. It becomes more like ours. They used to have an almost integral bloc, a geographical unit, with a single Iron Curtain separating their side from the rest of the world. One could almost draw a closed curved line on a globe with everything inside it Soviet bloc and everything outside it not. Yugoslavia was the only ambiguity. It in turn made little Albania an anomaly—only a small one, but its political detachment in the early 1960s confirms the point. Cuba has been the same problem magnified. "Blocness" no longer means what it did. In a geographically tight bloc, satellites can have degrees of affiliation with the U.S.S.R. without necessarily spoiling the definition of the "bloc." Distant satellites, though, not only can be more independent because of Soviet difficulty in imposing its will by violence but they further disturb the geographical neatness of the bloc. "Blocness" ceases to be all-or-none; it becomes a matter of degree.

This process can then infect the territories contiguous with the U.S.S.R. And if the Soviet Union tempers its deterrent threats, hedging on the distant countries or on countries not fully integrated, it invites examination of the credibility of its threats everywhere. Certain things like honor and outrage are not meant to be matters of degree. One can say that his homeland is inviolate only if he knows exactly what he means by "homeland" and it is not cluttered up with full-fledged states, protectorates, territories, and gradations of citizenship that make some places more "homeland" than others. Like virginity, the homeland wants an absolute definition. This character the Soviet bloc has been losing and may lose even more if it acquires a graduated structure like the old British Empire.

We credited the Soviets with effective deterrence and in doing so genuinely gave them some. We came at last to treat the Sino-Soviet split as a real one; but it would have been wiser not to have acknowledged their fusion in the first place. In our efforts to dramatize and magnify the Soviet threat, we sometimes present the Soviet Union with a deterrent asset of a kind that we find hard to create for ourselves. We should relieve the Soviets as much as we can of any obligation to respond to an American engagement with China as to an engagement with Soviet Russia itself. If we relieve the Soviets of the obligation, we somewhat undo their commitment. We should be trying to make North Vietnam seem much more remote from the Soviet bloc than Puerto Rico is from the United States, to keep China out of the category of Alaska, and not to concede to bloc countries a sense of immunity. Events may oblige us—some of these very countries may oblige us—to initiate some kind of military engagement in the future;[12] and we would be wise to decouple those areas, as much as we can, from Soviet military forces in advance.[13]

12. Events evidently caught up with this sentence!

13. Possibly the single greatest consequence of the nuclear test ban—and I see no evidence that it was intended in the West, or that it motivated the final negotiations —was to exacerbate the Sino-Soviet dispute on security policy and bring its military implications into the open. What a diplomatic coup it would have been, had it been contrived that way!

Escaping Commitments

Sometimes a country wants to get out of a commitment—to decouple itself. It is not easy. We may have regretted our commitment to Quemoy in 1958, but there was no graceful way to undo it at that time. The Berlin wall was a genuine embarrassment. We apparently had not enough of a commitment to feel obliged to use violence against the Berlin wall. We had undeniably some commitment; there was some expectation that we might take action and some belief that we ought to. We did not, and it cost us something. If nobody had ever expected us to do anything about the wall—if we had never appeared to have any obligation to prevent things like the wall, and if we had never made any claims about East Berlin that seemed inconsistent with the wall—the wall would have embarrassed us less. Some people on our side were disappointed when we let the wall go up. The United States government would undoubtedly have preferred not to incur that disappointment. Diplomatic statements about the character of our rights and obligations in East Berlin were an effort to dismantle any commitment we might previously have had. The statements were not fully persuasive. Had the United States government known all along that something like the wall might go up, and had it planned all along not to oppose it, diplomatic preparation might have made the wall less of an embarrassment. In this case there appeared to be some residual commitment that we had not honored, and we had to argue retroactively that our essential rights had not been violated and that nothing rightfully ours had been taken from us.

The Soviets had a similar problem over Cuba. Less than six weeks before the President's missile crisis address of October 22, 1962, the Soviet government had issued a formal statement about Cuba. "We have said and do repeat that if war is unleashed, if the aggressor makes an attack on one state or another and this state asks for assistance, the Soviet Union has the possibility from its own territory to render assistance to any peace-loving state and not only to Cuba. And let no one doubt that the Soviet Union will render such assistance." And further,

"The Soviet government would like to draw attention to the fact that one cannot now attack Cuba and expect that the aggressor will be free from punishment for this attack. If this attack is made, this will be the beginning of the unleashing of war." It was a long, argumentative statement, however, and acknowledged that "only a madman can think now that a war started by him will be a calamity only for the people against which it is unleashed." And the most threatening language was not singled out for solemn treatment but went along as part of the argument. So there was at least a degree of ambiguity.

President Kennedy's television broadcast of October 22 was directly aimed at the Soviet Union. It was so directly aimed that one can infer only a conscious decision to make this not a Caribbean affair but an East–West affair. It concerned Soviet missiles and Soviet duplicity, a Soviet challenge; and the President even went out of his way to express concern for the Cubans, his desire that they not be hurt, and his regret for the "foreign domination" that was responsible for their predicament. The President did not say that we had a problem with Cuba and hoped the Soviets would keep out of it; he said we had an altercation with the Soviet Union and hoped Cubans would not be hurt.

The Soviet statement the following day, circulated to the Security Council of the United Nations, was evidently an effort to structure the situation a little differently. It accused the United States of piracy on the high seas and of "trying to dictate to Cuba what policy it must pursue." It said the United States government was "assuming the right to demand that states should account to it for the way in which they organize their defense, and should notify it of what their ships are carrying on the high seas. The Soviet government firmly repudiates such claims." The statement also said, "Today as never before, statesmen must show calm and prudence, and must not countenance the rattling of weapons."And indeed there was no rattling of weapons in the Soviet statement. The most they said was, "The presence of powerful weapons, including nuclear rocket weapons, in the Soviet Union is acknowledged by all the peoples in the world to be the decisive factor in deterring the aggressive forces of imperialism from unleashing a world war of annihilation. This mission the Soviet Union will continue to discharge with all firmness and consistency." But "if the aggressors unleash war, the Soviet Union will inflict the most powerful blow in response." By implication, what the United States Navy was doing, or even might do, was piracy so far, and not war, and the "peace-loving states cannot but protest." [14]

The orientation was toward an American affront to Cuba, not a Soviet-American confrontation. The key American demand for the "prompt dismantling and withdrawal of all offensive weapons in Cuba" before the quarantine could be lifted—that is, the direct relation of President Kennedy's action to the Soviet missiles—was not directly addressed. The Soviets chose not to enhance their commitment to Cuba by construing the United States action as one obliging a firm Soviet response; they construed it as a Caribbean issue. Their language seemed designed to dismantle an incomplete commitment rather than to bolster it.

But just as one cannot incur a genuine commitment by purely verbal means, one cannot get out of it with cheap words either. Secretary Dulles in 1958 could not have said, "Quemoy? Who cares about Quemoy? It's not worth fighting over, and our defense perimeter will be neater without it." The United States never did talk its way cleanly out of the Berlin wall business. Even if the letter of our obligations was never violated, there are bound to be some who think the spirit demanded more. We had little obligation to intervene in Hungary in 1956, and the Suez crisis confused and screened it. Nevertheless, there was a possibility that the West might do something and it did not. Maybe this was a convenience, clarifying an implicit understanding between East and West. But the cost was not zero.

If commitments could be undone by declaration they would be worthless in the first place. The whole purpose of verbal or

14. David L. Larson, ed., *The "Cuban Crisis" of 1962, Selected Documents and Chronology* (Boston, Houghton Mifflin, 1963), pp. 7–17, 41–46, 50–54.

ritualistic commitments, of political and diplomatic commitments, of efforts to attach honor and reputation to a commitment, is to make the commitment manifestly hard to get out of
on short notice. Even the commitments not deliberately incurred, and the commitments that embarrass one in unforeseen
circumstances, cannot be undone cheaply. The cost is the
discrediting of other commitments that one would still like to be
credited.[15]

If a country does want to get off the hook, to get out of a
commitment deliberately incurred or one that grew up unintended, the opponent's cooperation can make a difference. The
Chinese Communists seemed not to be trying, from 1958 on, to
make it easy for the United States to decouple itself from
Quemoy. They maintained, and occasionally intensified, enough
military pressure on the island to make graceful withdrawal
difficult, to make withdrawal look like retreat under duress. It is
hard to escape the judgment that they enjoyed American discomfort over Quemoy, their own ability to stir things up at will
but to keep crises under their control, and their opportunity to
aggravate American differences with Chiang Kai-shek.

Circumventing an Adversary's Commitments

"Salami tactics," we can be sure, were invented by a child;
whoever first expounded the adult version had already understood the principle when he was small. Tell a child not to go in
the water and he'll sit on the bank and submerge his bare feet;

15. The most eloquent rebuff I have come across is the answer the Romans
received from the Volciani in Spain, whom they tried to unite with other Spanish
cities against Carthage shortly after Rome had declined to defend the allied Spanish
town of Saguntum against Hannibal and it had been terribly destroyed. "Men of
Rome," said the eldest among them, "it seems hardly decent to ask us to prefer
your friendship to that of Carthage, considering the precedent of those who have
been rash enough to do so. Was not your betrayal of your friends in Saguntum
even more brutal than their destruction by their enemies the Carthaginians? I suggest
you look for allies in some spot where what happened to Saguntum has never
been heard of. The fall of that town will be a signal and melancholy warning
to the peoples of Spain never to count upon Roman friendship nor to trust Rome's
word." *The War With Hannibal,* Aubrey de Selincourt, transl. (Baltimore, Penguin
Books, 1965), p. 43.

he is not yet "in" the water. Acquiesce, and he'll stand up; no more of him is in the water than before. Think it over, and he'll start wading, not going any deeper; take a moment to decide whether this is different and he'll go a little deeper, arguing that since he goes back and forth it all averages out. Pretty soon we are calling to him not to swim out of sight, wondering whatever happened to all our discipline.

Most commitments are ultimately ambiguous in detail. Sometimes they are purposely so, as when President Eisenhower and Secretary Dulles announced that an attack on Quemoy might or might not trigger an American response under the "Formosa Doctrine" according to whether or not it was interpreted as part of an assault, or prelude to an assault, on Formosa itself. Even more commitments are ambiguous because of the plain impossibility of defining them in exact detail. There are areas of doubt even in the most carefully drafted statutes and contracts; and even people who most jealously guard their rights and privileges have been known to settle out of court, to excuse an honest mistake, or to overlook a minor transgression because of the high cost of litigation. No matter how inviolate our commitment to some border, we are unlikely to start a war the first time a few drunken soldiers from the other side wander across the line and "invade" our territory. And there is always the possibility that some East German functionary on the Autobahn really did not get the word, or his vehicle really did break down in our lane of traffic. There is some threshold below which the commitment is just not operative, and even that threshold itself is usually unclear.

From this arises the low-level incident or probe, and tactics of erosion. One tests the seriousness of a commitment by probing it in a noncommittal way, pretending the trespass was inadvertent or unauthorized if one meets resistance, both to forestall the reaction and to avoid backing down. One stops a convoy or overflies a border, pretending the incident was accidental or unauthorized; but if there is no challenge, one continues or enlarges the operation, setting a precedent, establishing rights of thoroughfare or squatters' rights, pushing the commitment back or raising the threshold. The use of "volunteers" by Soviet countries to intervene in trouble spots was usually an effort to sneak under the fence rather than climb over it, not quite invoking the commitment, but simultaneously making the commitment appear porous and infirm. And if there is no sharp qualitative division between a minor transgression and a major affront, but a continuous gradation of activity, one can begin his intrusion on a scale too small to provoke a reaction, and increase it by imperceptible degrees, never quite presenting a sudden, dramatic challenge that would invoke the committed response. Small violations of a truce agreement, for example, become larger and larger, and the day never comes when the camel's back breaks under a single straw.

The Soviets played this game in Cuba for a long time, apparently unaware that the camel's back in that case could stand only a finite weight (or hoping the camel would get stronger and stronger as he got used to the weight). The Korean War may have begun as a low-level incident that was hoped to be beneath the American threshold of response, and the initial American responses (before the introduction of ground troops) may have been misjudged. Salami tactics do not always work. The uncertainty in a commitment often invites a low-level or noncommittal challenge; but uncertainty can work both ways. If the committed country has a reputation for sometimes, unpredictably, reacting where it need not, and not always collaborating to minimize embarrassment, loopholes may be less inviting. If one cannot get a reputation for always honoring commitments in detail, because the details are ambiguous, it may help to get a reputation for being occasionally unreasonable. If one cannot buy clearly identifiable and fully reliable trip-wires, an occasional booby trap placed at random may serve somewhat the same purpose in the long run.

Landlords rarely evict tenants by strong-arm methods. They have learned that steady cumulative pressures work just as well, though more slowly, and avoid provoking a violent response. It is far better to turn off the water and the electricity, and let the tenant suffer the cumulative pressure of unflushed toilets and

THE ART OF COMMITMENT

candles at night and get out voluntarily, than to start manhandling his family and his household goods. Blockade works slowly; it puts the decision up to the other side. To invade Berlin or Cuba is a sudden identifiable action, of an intensity that demands response; but to cut off supplies does little the first day and not much more the second; nobody dies or gets hurt from the initial effects of a blockade. A blockade is comparatively passive; the eventual damage results as much from the obstinacy of the blockaded territory as from the persistence of the blockading power. And there is no well-defined moment before which the blockading power may quail, for fear of causing the ultimate collapse.

President Truman appreciated the value of this tactic in June 1945. French forces under de Gaulle's leadership had occupied a province in Northern Italy, contrary to Allied plans and American policy. They announced that any effort of their allies to dislodge them would be treated as a hostile act. The French intended to annex the area as a "minor frontier adjustment." It would have been extraordinarily disruptive of Allied unity, of course, to expel the French by force of arms; arguments got nowhere, so President Truman notified de Gaulle that no more supplies would be issued to the French army until it had withdrawn from the Aosta Valley. The French were absolutely dependent on American supplies and the message brought results. This was "nonhostile" pressure, not quite capable of provoking a militant response, therefore safe to use (and effective). A given amount of coercive pressure exercised over an extended period of time, allowed to accumulate its own momentum, is a common and effective technique of bypassing somebody's commitment.

The Distinction Between Deterrence and "Compellence"

Blockade illustrates the typical difference between a threat intended to make an adversary do something and a threat intended to keep him from starting something. The distinction is in the timing and in the initiative, in who has to make the first move, in whose initiative is put to the test. To deter an enemy's advance it may be enough to burn the escape bridges behind me, or to rig a trip-wire between us that automatically blows us both up when he advances. To *compel* an enemy's retreat, though, by some threat of engagement, I have to be committed to *move*. (This requires setting fire to the grass behind me as I face the enemy, with the wind blowing toward the enemy.) I can block your car by placing mine in the way; my *deterrent* threat is passive, the decision to collide is up to you. But if you find me in your way and threaten to collide unless I move, you enjoy no such advantage; the decision to collide is still yours, and I still enjoy deterrence. You have to arrange to *have* to collide unless I move, and that is a degree more complicated. You have to get up so much speed that you cannot stop in time and that only I can avert the collision; this may not be easy. If it takes more time to start a car than to stop one, you may be unable to give me the "last clear chance" to avoid collision by vacating the street.

The threat that compels rather than deters often requires that the punishment be administered *until* the other acts, rather than *if* he acts. This is because often the only way to become committed to an action is to initiate it. This means, though, that the action initiated has to be tolerable to the initiator, and tolerable over whatever period of time is required for the pressure to work on the other side. For deterrence, the trip-wire can threaten to blow things up out of all proportion to what is being protected, because if the threat works the thing never goes off. But to hold a large bomb and threaten to throw it *unless* somebody moves cannot work so well; the threat is not believable until the bomb is actually thrown and by then the damage is done.[16]

There is, then, a difference between *deterrence* and what we

16. A nice illustration occurs in the movie version of *A High Wind in Jamaica*. The pirate captain, Chavez, wants his captive to tell where the money is hidden and puts his knife to the man's throat to make him talk. After a moment or two, during which the victim keeps his mouth shut, the mate laughs. "If you cut his throat he can't tell you. He knows it. And he knows you know it." Chavez puts his knife away and tries something else.

might, for want of a better word, call *compellence*. The dictionary's definition of "deter" corresponds to contemporary usage: to turn aside or discourage through fear; hence, to prevent from action by fear of consequences. A difficulty with our being an unaggressive nation, one whose announced aim has usually been to contain rather than to roll back, is that we have not settled on any conventional terminology for the more active kind of threat. We have come to use "defense" as a euphemism for "military," and have a Defense Department, a defense budget, a defense program, and a defense establishment; if we need the other word, though, the English language provides it easily. It is "offense." We have no such obvious counterpart to "deterrence." "Coercion" covers the meaning but unfortunately includes "deterrent" as well as "compellent" intentions. "Intimidation" is insufficiently focused on the particular behavior desired. "Compulsion" is all right but its adjective is "compulsive," and that has come to carry quite a different meaning. "Compellence" is the best I can do.[17]

Deterrence and compellence differ in a number of respects, most of them corresponding to something like the difference between statics and dynamics. Deterrence involves setting the stage—by announcement, by rigging the trip-wire, by incurring the obligation—and *waiting*. The overt act is up to the opponent. The stage-setting can often be nonintrusive, nonhostile,

17. J. David Singer has used a nice pair of nouns, "persuasion" and "dissuasion," to make the same distinction. It is the adjectives that cause trouble; "persuasive" is bound to suggest the adequacy or credibility of a threat, not the character of its objective. Furthermore, "deterrent" is here to stay, at least in the English language. Singer's breakdown goes beyond these two words and is a useful one; he distinguishes whether the subject is desired to *act* or *abstain,* whether or not he is *presently* acting or abstaining, and whether he is likely (in the absence of threats and offers) to *go on* acting or abstaining. (If he is behaving, and is likely—but not certain—to go on behaving, there can still be reason to "reinforce" his motivation to behave.) Singer distinguishes also "rewards" and "penalties" as well as threats and offers; while the rewards and "penalties" can be the *consequences* of threats and offers, they can also be *gratuitous,* helping to communicate persuasively some new and continuing threat or offer. See his article, "Inter-Nation Influence: A Formal Model," *American Political Science Review, 17* (1963), 420–30.

nonprovocative. The act that is intrusive, hostile, or provocative is usually the one to be deterred; the deterrent threat only changes the consequences *if* the act in question—the one to be deterred—is then taken. Compellence, in contrast, usually involves *initiating* an action (or an irrevocable commitment to action) that can cease, or become harmless, only if the opponent responds. The overt act, the first step, is up to the side that makes the compellent threat. To deter, one digs in, or lays a minefield, and waits—in the interest of inaction. To compel, one gets up enough momentum (figuratively, but sometimes literally) to make the other *act* to avoid collision.

Deterrence tends to be indefinite in its timing. "If you cross the line we shoot in self-defense, or the mines explode." When? Whenever you cross the line—preferably never, but the timing is up to you. If *you* cross it, *then* is when the threat is fulfilled, either automatically, if we've rigged it so, or by obligation that immediately becomes due. But we can wait—preferably forever; that's our purpose.

Compellence has to be definite: We move, and you must get out of the way. By when? There has to be a deadline, otherwise tomorrow never comes. If the action carries no deadline it is only a posture, or a ceremony with no consequences. If the compellent advance is like Zeno's tortoise that takes infinitely long to reach the border by traversing, with infinite patience, the infinitely small remaining distances that separate him from collision, it creates no inducement to vacate the border. Compellence, to be effective, can't wait forever. Still, it has to wait a little; collision can't be instantaneous. The compellent threat has to be put in motion to be credible, and *then* the victim must yield. Too little time, and compliance becomes impossible; too much time, and compliance becomes unnecessary. Thus compellence involves timing in a way that deterrence typically does not.

In addition to the question of "when," compellence usually involves questions of where, what, and how much. "Do nothing" is simple, "Do something" ambiguous. "Stop where you are" is simple; "Go back" leads to "How far?" "Leave me alone" is simple; "Cooperate" is inexact and open-ended. A deterrent position—a status quo, in territory or in more figurative terms—can often be surveyed and noted; a compellent advance has to be *projected* as to destination, and the destination can be unclear in intent as well as in momentum and braking power. In a deterrent threat, the objective is often communicated by the very preparations that make the threat credible; the trip-wire often demarcates the forbidden territory. There is usually an inherent connection between *what* is threatened and what it is threatened *about*. Compellent threats tend to communicate only the general direction of compliance, and are less likely to be self-limiting, less likely to communicate in the very design of the threat just what, or how much, is demanded. The garrison in West Berlin can hardly be misunderstood about what it is committed to resist; if it ever intruded into East Berlin, though, to induce Soviet or German Democratic Republic forces to give way, there would be no such obvious interpretation of where and how much to give way unless the adventure could be invested with some unmistakable goal or limitation— a possibility not easily realized.

The Quemoy escapade is again a good example: Chiang's troops, once on the island, especially if evacuation under fire appeared infeasible, had the static clarity that goes with commitment to an indefinite status quo, while the commitment just to *send* troops to defend it (or air and naval support) according to whether a Communist attack there was or was not prelude to an attack on Formosa lacked that persuasive quality, reminding us that though deterrent threats tend to have the advantages mentioned above they do not always achieve them. (The ambiguous case of Quemoy actually displays the compellent ambiguity, seen in reverse: a "compellent" Communist move against Quemoy was to be accommodated, as long as its extent could be reliably projected to a terminus short of Formosa; if the Communists thought we meant it, it was up to them to design an action that visibly embodied that limitation.) An American or NATO action to relieve Budapest in 1956—without major engagement but in the hope the Soviets would give way rather than fight—would have had the dynamic quality of "compellence" in contrast to Berlin: the stopping point would have been a variable, not a constant. Even "Budapest" would have needed a definition, and might have become all of Hungary—and after Hungary, what?—if the Soviets initially gave way. The enterprise might have been designed to embody its specific intent, but it would have taken a lot of designing backed up by verbal assurances.

Actually, any coercive threat requires corresponding *assurances;* the object of a threat is to give somebody a choice. To say, "One more step and I shoot," can be a deterrent threat only if accompanied by the implicit assurance, "And if you stop I won't." Giving notice of *unconditional* intent to shoot gives him no choice (unless by behaving as we wish him to behave the opponent puts himself out of range, in which case the effective threat is, "Come closer and my fire will kill you, stay back and it won't"). What was said above about deterrent threats being typically less ambiguous in intent can be restated: the corresponding assurances—the ones that, together with the threatened response, define the opponent's choice—are clearer than those that can usually be embodied in a compellent action. (Ordinary blackmailers, not just nuclear, find the "assurances" troublesome when their threats are compellent.) [18]

They are, furthermore, confirmed and demonstrated over time; as long as he stays back, and we don't shoot, we fulfill the assurances and confirm them. The assurances that accompany a compellent action—move back a mile and I won't shoot (otherwise I shall) and I won't then try again for a second mile—are

18. The critical role of "assurances" in completing the structure of a threat, in making the threatened consequences persuasively *conditional* on behavior so that the victim is offered a choice, shows up in the offers of amnesty, safe passage, or forgiveness that must often be made credible in inducing the surrender of rebels or the capitulation of strikers or protesters. Even libraries and internal revenue agencies depend on parallel offers of forgiveness when they embark on campaigns to coerce the return of books or payment of back taxes. In personal life I have sometimes relied, like King Lear, on the vague threat that my wrath will be aroused (with who knows what awful consequences) if good behavior is not forthcoming, making a tentative impression on one child, only to have the threat utterly nullified by another's pointing out that "Daddy's mad already."

harder to demonstrate in advance, unless it be through a long past record of abiding by one's own verbal assurances.

Because in the West we deal mainly in deterrence, not compellence, and deterrent threats tend to convey their assurances implicitly, we often forget that *both* sides of the choice, the threatened penalty and the proffered avoidance or reward, need to be credible. The need for assurances—not just verbal but fully credible—emerges clearly as part of "deterrence" in discussions of surprise attack and "preemptive war." An enemy belief that we are about to attack anyway, not after he does but possibly before, merely raises his incentive to do what we wanted to deter and to do it even more quickly. When we do engage in compellence, as in the Cuban crisis or in punitive attacks on North Vietnam that are intended to make the North Vietnamese government act affirmatively, the assurances are a critical part of the definition of the compellent threat.

One may deliberately choose to be unclear and to keep the enemy guessing either to keep his defenses less prepared or to enhance his anxiety. But if one wants not to leave him in doubt about what will satisfy us, we have to find credible ways of communicating, and communicating both what we want and what we do not want. There is a tendency to emphasize the communication of what we shall *do* if he misbehaves and to give too little emphasis to communicating *what* behavior will satisfy us. Again, this is natural when deterrence is our business, because the prohibited misbehavior is often approximately defined in the threatened response; but when we must start something that then has to be stopped, as in compellent actions, it is both harder and more important to know our aims and to communicate. It is particularly hard because the mere initiation of an energetic coercive campaign, designed for compellence, disturbs the situation, leads to surprises, and provides opportunity and temptation to reexamine our aims and change them in mid course. Deterrence, if wholly successful, can often afford to concentrate on the initiating events—what happens *next* if he misbehaves. Compellence, to be successful, involves an action that must be brought to successful closure. The payoff comes at the end, as does the disaster if the project fails.

The compellent action will have a time schedule of its own, and unless it is carefully chosen it may not be reconcilable with the demands that are attached to it. We cannot usefully threaten to bomb Cuba next Thursday unless the Russians are out by next month, or conduct a six weeks' bombing campaign in North Vietnam and stop it when the Vietcong have been quiescent for six months. There will be limits, probably, to how long the compellent action can be sustained without costing or risking too much, or exhausting itself or the opponent so that he has nothing left to lose. If it cannot induce compliance within that time—and this depends on whether compliance is physically or administratively feasible within that time—it cannot accomplish anything (unless the objective was only an excuse for some act of conquest or punishment). The compellent action has to be one that can be stopped or reversed when the enemy complies, or else there is no inducement.

If the opponent's compliance necessarily takes time—if it is sustained good behavior, cessation of an activity that he must not resume, evacuation of a place he must not reenter, payment of tribute over an extended period, or some constructive activity that takes time to accomplish—the compellent threat requires some commitment, pledge, or guarantee, or some hostage, or else must be susceptible of being resumed or repeated itself. Particularly in a crisis, a Cuban crisis or a Vietnamese crisis, there is strong incentive to get compliance quickly to limit the risk or damage. Just finding conditions that can be met on the demanding time schedule of a dangerous crisis is not easy. The ultimate demands, the objectives that the compellent threat is really aimed at, may have to be achieved indirectly, by taking pledges or hostages that can be used to coerce compliance after the pressure has been relieved.[19] Of course, if some kind of

19. Lord Portal's account of the coercive bombing of the villages of recalcitrant Arab tribesmen (after warning to permit evacuation) includes the terms that were demanded. Among them were hostages—literal hostages, people—as well as a fine; otherwise the demand was essentially cessation of the raids or other misbehavior that had brought on the bombing. The hostages were apparently partly to permit subsequent enforcement without repeated bombing, partly to symbolize, together with the fine, the tribe's intent to comply. See Portal, "Air Force Cooperation in Policing the Empire," pp. 343–58.

surrender statement or acknowledgement of submission, some symbolic knuckling under, will itself achieve the object, verbal compliance may be enough. It is inherent in an intense crisis that the conditions for bringing it to a close have to be of a kind that can be met quickly; that is what we mean by an "intense crisis," one that compresses risk, pain, or cost into a short span of time or that involves actions that cannot be sustained indefinitely. If we change our compellent threat from slow pressure to intense, we have to change our demands to make them fit the urgent timing of a crisis.

Notice that to deter *continuance* of something the opponent is already doing—harassment, overflight, blockade, occupation of some island or territory, electronic disturbance, subversive activity, holding prisoners, or whatever it may be—has some of the character of a compellent threat. This is especially so of the timing, of who has to take the initiative. In the more static case we want him to go on *not* doing something; in this more dynamic case we want him to *change* his behavior. The "when" problem arises in compelling him to stop, and the compellent action may have to be initiated, not held in waiting like the deterrent threat. The problems of "how much" may not arise if it is some discrete, well defined activity. "At all" may be the obvious answer. For U-2 flights or fishing within a twelve-mile limit, that may be the case; for subversive activity or support to insurgents, "at all" may itself be ambiguous because the activity is complex, ill defined, and hard to observe or attribute.

Blockade, harassment, and "salami tactics" can be interpreted as ways of evading the dangers and difficulties of compellence. Blockade in a cold war sets up a tactical "status quo" that is damaging in the long run but momentarily safe for both sides unless the victim tries to run the blockade. President Kennedy's overt act of sending the fleet to sea, in "quarantine" of Cuba in October 1962, had some of the quality of deterrent "stage setting"; the Soviet government then had about forty-eight hours to instruct its steamers whether or not to seek collision. Low-level intrusion, as discussed earlier, can be a way of letting the opponent turn his head and yield a little, or it can be a way of starting a compellent action in low gear, without the conviction that goes with greater momentum but also without the greater risk. Instead of speeding out of control toward our car that blocks his way, risking our inability to see him and get our engines started in time to clear his path, he approaches slowly and nudges fenders, crushing a few lights and cracking some paint. If we yield he can keep it up, if not he can cut his losses. And if he makes it look accidental, or can blame it on an impetuous chauffeur, he may not even lose countenance in the unsuccessful try.

Defense and Deterrence, Offense and Compellence

The observation that deterrent threats are often passive, while compellent threats often have to be active, should not be pressed too far. Sometimes a deterrent threat cannot be made credible in advance, and the threat has to be made lively when the prohibited action is undertaken. This is where *defense* and *deterrence* may merge, forcible defense being undertaken in the hope, perhaps with the main purpose, of demonstrating by resistance that the conquest will be costly, even if successful, too costly to be worthwhile. The idea of "graduated deterrence" and much of the argument for a conventional warfare capability in Europe are based on the notion that if passive deterrence initially fails, the more active kind may yet work. If the enemy act to be deterred is a once-for-all action, incapable of withdrawal, rather than progressive over time, any failure of deterrence is complete and final; there is no second chance. But if the aggressive move takes time, if the adversary did not believe he would meet resistance or did not appreciate how costly it would be, one can still hope to demonstrate that the threat is in force, after he begins. If he expected no opposition, encountering some may cause him to change his mind.

There is still a distinction here between forcible defense and defensive action intended to deter. If the object, and the only hope, is to resist successfully, so that the enemy *cannot* succeed even if he tries, we can call it pure defense. If the object is to *induce* him not to proceed, by making his encroachment painful or costly, we can call it a "coercive" or "deterrent" defense. The language is clumsy but the distinction is valid. Resistance that might otherwise seem futile can be worthwhile if, though incapable of blocking aggression, it can nevertheless threaten to make the cost too high. This is "active" or "dynamic" deterrence, deterrence in which the threat is communicated by progressive fulfillment. At the other extreme is forcible defense with good prospect of blocking the opponent but little promise of hurting; this would be purely defensive.

Defensive action may even be undertaken with no serious hope of repelling or deterring enemy action but with a view to making a "successful" conquest costly enough to deter repetition by the same opponent or anyone else. This is of course the rationale for reprisals after the fact; they cannot undo the deed but can make the books show a net loss and reduce the incentive next time. Defense can sometimes get the same point across, as the Swiss demonstrated in the fifteenth century by the manner in which they lost battles as well as by the way they sometimes won them. "The [Swiss] Confederates were able to reckon their reputation for obstinate and invincible courage as one of the chief causes which gave them political importance. . . . It was no light matter to engage with an enemy who would not retire before any superiority in numbers, who was always ready for a fight, who would neither give nor take quarter." [20] The Finns demonstrated five hundred years later that the principle still works. The value of local resistance is not measured solely by local success. This idea of what we might call "punitive resistance" could have been part of the rationale for the American commitment of forces in Vietnam.[21]

"Compellence" is more like "offense." *Forcible* offense is taking something, occupying a place, or disarming an enemy or a territory, by some direct action that the enemy is unable to block. "Compellence" is *inducing* his withdrawal, or his ac-

20. C.W.C. Oman, *The Art of War in the Middle Ages* (Ithaca, Cornell University Press, 1953), p. 96.

21. An alternative, but not inconsistent, treatment of some of these distinctions is in Glenn H. Snyder, *Deterrence and Defense* (Princeton, Princeton University Press, 1961), pp. 5–7, 9–16, 24–40.

quiescence, or his collaboration by an action that threatens to
hurt, often one that could not forcibly accomplish its aim but
that, nevertheless, can hurt enough to induce compliance. The
forcible and the coercive are both present in a campaign that
could reach its goal against resistance, and would be worth the
cost, but whose cost is nevertheless high enough so that one
hopes to induce compliance, or to deter resistance, by making
evident the intent to proceed. Forcible action, as mentioned in
Chapter 1, is limited to what can be accomplished without
enemy collaboration; compellent threats can try to induce more
affirmative action, including the exercise of authority by an
enemy to bring about the desired results.

War itself, then, can have deterrent or compellent intent, just
as it can have defensive or offensive aims. A war in which both
sides can hurt each other but neither can forcibly accomplish its
purpose could be compellent on one side, deterrent on the other.
Once an engagement starts, though, the difference between
deterrence and compellence, like the difference between defense and offense, may disappear. There can be legal and moral
reasons, as well as historical reasons, for recalling the status quo
ante; but if territory is in dispute, the strategies for taking it,
holding it, or recovering it may not much differ as between the
side that originally possessed it and the side that coveted it, once
the situation has become fluid. (In a local tactical sense,
American forces were often on the "defensive" in North Korea
and on the "offensive" in South Korea.) The coercive aspect of
warfare may be equally compellent on both sides, the only
difference perhaps being that the demands of the defender, the
one who originally possessed what is in dispute, may be clearly
defined by the original boundaries, whereas the aggressor's
demands may have no such obvious definition.

The Cuban crisis is a good illustration of the fluidity that sets
in once passive deterrence has failed. The United States made
verbal threats against the installation of weapons in Cuba but
apparently some part of the threat was unclear or lacked
credibility and it was transgressed. The threat lacked the automaticity that would make it fully credible, and without some

automaticity it may not be clear to either side just where the threshold is. Nor was it physically easy to begin moderate resistance after the Russians had crossed the line, and to increase the resistance progressively to show that the United States meant it. By the time the President determined to resist, he was no longer in a deterrent position and had to embark on the more complicated business of compellence. The Russian missiles could sit waiting, and so could Cuban defense forces; the next overt act was up to the President. The problem was to prove to the Russians that a potentially dangerous action was forthcoming, without any confidence that verbal threats would be persuasive and without any desire to initiate some irreversible process just to prove, to everybody's grief, that the United States meant what it said.

The problem was to find some action that would communicate the threat, an action that would promise damage if the Russians did not comply but minimum damage if they complied quickly enough, and an action that involved enough momentum or commitment to put the next move clearly up to the Russians. Any overt act against a well-defended island would be abrupt and dramatic; various alternatives were apparently considered, and in the end an action was devised that had many of the virtues of static deterrence. A blockade was thrown around the island, a blockade that by itself could not make the missiles go away. The blockade did, however, threaten a minor military confrontation with major diplomatic stakes—an encounter between American naval vessels and Soviet merchant ships bound for Cuba. Once in place, the Navy was in a position to wait; it was up to the Russians to decide whether to continue. If Soviet ships had been beyond recall, the blockade would have been a preparation for inevitable engagement; with modern communications the ships were not beyond recall, and the Russians were given the last clear chance to turn aside. Physically the Navy could have avoided an encounter; diplomatically, the declaration of quarantine and the dispatch of the Navy meant that American evasion of the encounter was virtually out of the question. For the Russians, the diplomatic cost of turning freighters around, or even letting one be examined, proved not to be prohibitive.

Thus an initial deterrent threat failed, a compellent threat was called for, and by good fortune one could be found that had some of the static qualities of a deterrent threat.[22]

There is another characteristic of compellent threats, arising in the need for affirmative action, that often distinguishes them from deterrent threats. It is that the very act of compliance—of doing what is demanded—is more conspicuously compliant, more recognizable as submission under duress, than when an act is merely withheld in the face of a deterrent threat. Compliance is likely to be less casual, less capable of being rationalized as something that one was going to do anyhow. The Chinese did not need to acknowledge that they shied away from Quemoy or Formosa because of American threats, and the Russians need not have agreed that it was NATO that deterred them from conquering Western Europe, and no one can be sure. Indeed, if a deterrent threat is created before the proscribed act is even contemplated, there need never be an explicit decision *not* to transgress, just an absence of any temptation *to* do the thing prohibited. The Chinese still say they will take Quemoy in their own good time; and the Russians go on saying that their intentions against Western Europe were never aggressive.

The Russians cannot, though, claim that they were on the point of removing their missiles from Cuba anyway, and that the President's television broadcast, the naval quarantine and

22. Arnold Horelick agrees with this description. "As an initial response the quarantine was considerably less than a direct application of violence, but considerably more than a mere protest or verbal threat. The U.S. Navy placed itself physically between Cuba and Soviet ships bound for Cuban ports. Technically, it might still have been necessary for the United States to fire the first shot had Khrushchev chosen to defy the quarantine, though other means of preventing Soviet penetration might have been employed. But once the quarantine was effectively established— which was done with great speed—it was Khrushchev who had to make the next key decision: whether or not to risk releasing the trip-wire." "The Cuban Missile Crisis," *World Politics, 16* (1964), 385. This article and the Adelphi Paper of Albert and Roberta Wohlstetter mentioned in an earlier note are the best strategic evaluations of the Cuban affair that I have discovered.

threats of more violent action, had no effect.[23] If the North Vietnamese dramatically issue a call to the Vietcong to cease activity and to evacuate South Vietnam, it is a conspicuous act of submission. If the Americans had evacuated Guantanamo when Castro turned off the water, it would have been a conspicuous act of submission. If an earthquake or change in the weather had caused the water supply to dry up at Guantanamo, and if the Americans had found it wholly uneconomical to supply the base by tanker, they might have quit the place without seeming to submit to Castro's cleverness or seeming afraid to take reprisals against their ungracious host. Similarly, the mere act of bombing North Vietnam changed the status of any steps that the North Vietnamese might take to comply with American wishes. It can increase their desire, if the tactic is successful, to reduce support for the Vietcong; but it also increases the cost of doing so. Secretary Dulles used to say that while we had no vital interest in Quemoy we could not afford to evacuate under duress; intensified Chinese pressure always led to intensified determination to resist it.[24]

If the object is actually to impose humiliation, to force a showdown and to get an acknowledgement of submission, then the "challenge" that is often embodied in an active compellent threat is something to be exploited. President Kennedy undoubtedly wanted some conspicuous compliance by the Soviet Union during the Cuban missile crisis, if only to make clear to the Russians themselves that there were risks in testing how much the American government would absorb such ventures. In Vietnam the problem appeared the opposite; what was most

23. The tendency for affirmative action to appear compliant is vividly illustrated by the widespread suspicion—one that could not be effectively dispelled—that the U.S. missiles removed from Turkey in the wake of the Cuban crisis were part of a bargain, tacit if not explicit.

24. Almost everyone in America, surely including the President and the Secretary of State, would have been relieved in the late 1950s if an earthquake or volcanic action had caused Quemoy to sink slowly beneath the surface of the sea. Evacuation would then not have been retreat, and an unsought commitment that had proved peculiarly susceptible to Communist China's manipulation would have been disposed of. Such is the intrinsic value of some territories that have to be defended!

urgently desired was to reduce the support for the Vietcong from the North, and any tendency for the compellent pressure of bombing to produce a corresponding resistance would have been deprecated. But it cannot always be avoided, and if it cannot, the compellent threat defeats itself.

Skill is required to devise a compellent action that does not have this self-defeating quality. There is an argument here for sometimes not being too explicit or too open about precisely what is demanded, if the demands can be communicated more privately and noncommittally. President Johnson was widely criticized in the press, shortly after the bombing attacks began in early 1965, for not having made his objectives entirely clear. How could the North Vietnamese comply if they did not know exactly what was wanted? Whatever the reason for the American Administration's being somewhat inexplicit—whether it chose to be inexplicit, did not know how to be explicit, or in fact was explicit but only privately—an important possibility is that vague demands, though hard to understand, can be less embarrassing to comply with. If the President had to be so explicit that any European journalist knew exactly what he demanded, and if the demands were concrete enough to make compliance recognizable when it occurred, any compliance by the North Vietnamese regime would necessarily have been fully public, perhaps quite embarrassingly so. The action could not be hidden nor the motive so well disguised as if the demands were more privately communicated or left to inference by the North Vietnamese.

Another serious possibility is suggested by the North Vietnamese case: that the initiator of a compellent campaign is not himself altogether sure of what action he wants, or how the result that he wants can be brought about. In the Cuban missile case it was perfectly clear what the United States government wanted, clear that the Soviets had the ability to comply, fairly clear how quickly it could be done, and reasonably clear how compliance might be monitored and verified, though in the end there might be some dispute about whether the Russians had left behind things they were supposed to remove. In the Vietnamese case, we can suppose that the United States government did not know in detail just how much control or influence the North Vietnamese regime had over the Vietcong; and we can even suppose that the North Vietnamese regime itself might not have been altogether sure how much influence it would have in commanding a withdrawal or in sabotaging the movement that had received its moral and material support. The United States government may not have been altogether clear on which kinds of North Vietnamese help—logistical help, training facilities, sanctuary for the wounded, sanctuary for intelligence and planning activities, communications relay facilities, technical assistance, advisors and combat leaders in the field, political and doctrinal assistance, propaganda, moral support or anything else—were most effective and essential, or most able to be withdrawn on short notice with decisive effects. And possibly the North Vietnamese did not know. The American government may have been in the position of demanding *results* not specific *actions,* leaving it to the North Vietnamese through overt acts, or merely through reduced support and enthusiasm, to weaken the Vietcong or to let it lose strength. Not enough is known publicly to permit us to judge this Vietnamese instance; but it points up the important possibility that a compellent threat may have to be focused on results rather than contributory deeds, like the father's demand that his son's school grades be improved, or the extortionist's demand, "Get me money. I don't care how you get it, just get it." A difficulty, of course, is that results are more a matter of interpretations than deeds usually are. Whenever a recipient of foreign aid, for example, is told that it must eliminate domestic corruption, improve its balance of payments, or raise the quality of its civil service, the results tend to be uncertain, protracted, and hard to attribute. The country may try to comply and fail; with luck it may succeed without trying; it may have indifferent success that is hard to judge; in any case compliance is usually arguable and often visible only in retrospect.

Even more than deterrence, compellence requires that we recognize the difference between an individual and a government. To coerce an individual it may be enough to persuade him to change his mind; to coerce a government it may not be necessary, but it also may not be sufficient, to cause individuals to change their minds. What may be required is some change in the complexion of the government itself, in the authority, prestige, or bargaining power of particular individuals or factions or parties, some shift in executive or legislative leadership. The Japanese surrender of 1945 was marked as much by changes in the structure of authority and influence within the government as by changes in attitude on the part of individuals. The victims of coercion, or the individuals most sensitive to coercive threats, may not be directly in authority; or they may be hopelessly committed to non-compliant policies. They may have to bring bureaucratic skill or political pressure to bear on individuals who do exercise authority, or go through processes that shift authority or blame to others. In the extreme case governing authorities may be wholly unsusceptible to coercion—may, as a party or as individuals, have everything to lose and little to save by yielding to coercive threats—and actual revolt may be essential to the process of compliance, or sabotage or assassination. Hitler was uncoercible; some of his generals were not, but they lacked organization and skill and failed in their plot. For working out the incentive structure of a threat, its communication requirements and its mechanism, analogies with individuals are helpful; but they are counterproductive if they make us forget that a government does not reach a decision in the same way as an individual in a government. Collective decision depends on the internal politics and bureaucracy of government, on the chain of command and on lines of communication, on party structures and pressure groups, as well as on individual values and careers. This affects the speed of decision, too.

"Connectedness" in Compellent Threats

As mentioned earlier, a deterrent threat usually enjoys some *connectedness* between the proscribed action and the threatened response. The connection is sometimes a physical one, as when troops are put in Berlin to defend Berlin. Compellent actions often have a less well-defined connectedness; and the question arises whether they ought to be connected at all. If the object is to harass, to blockade, to scare or to inflict pain or damage until an adversary complies, why cannot the connection be made verbally? If the Russians want Pan-American Airways to stop using the air corridor to Berlin, why can they not harass the airline on its Pacific routes, announcing that harassment will continue until the airline stops flying to Berlin? When the Russians put missiles in Cuba, why cannot the President quarantine Vladivostok, stopping Soviet ships outside, say, a twelve-mile limit, or perhaps denying them access to the Suez or Panama Canal? And if the Russians had wanted to counter the President's quarantine of Cuba, why could they not blockade Norway?[25]

A hasty answer may be that it just is not done, or is not "justified," as though connectedness implied justice, or as though justice were required for effectiveness. Surely that is part of the answer; there is a legalistic or diplomatic, perhaps a casuistic, propensity to keep things connected, to keep the threat and the demand in the same currency, to do what seems reasonable. But why be reasonable, if results are what one

25. It has often been said that American tactical superiority and ease of access in the Caribbean (coupled with superiority in strategic weaponry) account for the success in inducing evacuation of the Soviet missiles. Surely that was crucial; but equally significant was the universal tendency—a psychological phenomenon, a tradition or convention shared by Russians and Americans—to *define* the conflict in Caribbean terms, not as a contest, say, in the blockade of each other's island allies, not as a counterpart of their position in Berlin, not as a war of harassment against strategic weapons outside national borders. The countermeasures and counterpressures available to the Russians might have looked very different to the "Russian" side if this had been a game on an abstract board rather than an event in historical time in a particular part of the real world. The Russians tried (as did some unhelpful Americans) to find a connection between Soviet missiles in Cuba and American missiles in Turkey, but the connection was evidently not persuasive enough for the Russians to be confident that, if the dispute led to military action or pressure against Turkey, *that* definition would hold and things would go no further. The Caribbean definition had more coherence or integrity than a Cuban–Turkish definition would have had, or, in terms of reciprocal blockade, a Cuban–United Kingdom definition would have had. The risk of further metastasis must have inhibited any urge to let the crisis break out of its original Caribbean definition.

wants? Habit, tradition, or some psychological compulsion may explain this connectedness, but it has to be asked whether they make it wise.

There are undoubtedly some good reasons for designing a compellent campaign that is connected with the compliance desired. One is that it helps to communicate the threat itself; it creates less uncertainty about what is demanded, what pressure will be kept up until the demands are complied with and then relaxed once they are. Actions not only speak louder than words on many occasions, but like words they can speak clearly or confusingly. To the extent that actions speak, it helps if they reinforce the message rather than confuse it.

Second, if the object is to induce compliance and not to start a spiral of reprisals and counteractions, it is helpful to show the limits to what one is demanding, and this can often be best shown by designing a campaign that distinguishes what is demanded from all the other objectives that one might have been seeking but is not. To harass aircraft in the Berlin air corridor communicates that polar flights are not at issue; to harass polar flights while saying that it is punishment for flying in the Berlin corridor does not so persuasively communicate that the harassment will stop when the Berlin flights stop, or that the Russians will not think of a few other favors they would like from the airline before they call off their campaign. Most of the problems of defining the threat and the demands that go with it, of offering assurance about what is not demanded and of promising cessation once compliance is forthcoming, are aggravated if there is no connection between the compellent action (or the threat of it) and the issue being bargained over.

The same question can arise with deterrent threats; sometimes they lack connectedness. To threaten the Chinese mainland in the event of an overland attack on India has a minimum of connectedness. If the threatened response is massive enough, though, it may seem to comprise or to include the local area and not merely to depart from it. But it often lacks some of the credibility, through automatic involvement, that can be achieved by connecting the response physically to the provocation itself. Contingent actions—not actions *initiated* to induce compliance, ut actions *threatened* against potential provocation—often need the credibility that connectedness can give them.

Connectedness in fact provides something of a scheme for classifying compellent threats and actions. The ideal compellent action would be one that, once initiated, causes minimal harm if compliance is forthcoming and great harm if compliance is not forthcoming, is consistent with the time schedule of feasible compliance, is beyond recall once initiated, and cannot be stopped by the party that started it but *automatically* stops upon compliance, with all this fully understood by the adversary. Only *he* can avert the consequences; he can do it only by complying; and compliance automatically precludes them. His is then the "last clear chance" to avert the harm or catastrophe; and it would not even matter which of the two most feared the consequences as long as the adversary knew that only he, by complying, could avert them. (Of course, whatever is demanded of him must be less unattractive to him than the threatened consequences, and the manner of threatened compliance must not entail costs in prestige, reputation, or self-respect that outweigh the threat.)

It is hard to find significant international events that have this perfectionist quality. There are situations, among cars on highways or in bureaucratic bargaining or domestic politics, where one comes across such ideal compellent threats; but they usually involve physical constraints or legal arrangements that tie the hand of the initiator in a way that is usually not possible in international relations. Still, if we include actions that the initiator can physically recall but not without intolerable cost, so that it is evident he would not go back even if it is equally clear that he could, we can find some instances. An armed convoy on a Berlin Autobahn may sometimes come close to having this quality.

A degree less satisfactory is the compellent action of which the consequences can be averted by *either* side, by the initiator's changing his mind just in time or by his adversary's compliance. Because he can stop before the consequences mount up, this type of compellent action may be less risky for the party that starts it; there is a means of escape, though it may become a test of nerves, or a test of endurance, each side hoping the other will back down, both sides possibly waiting too long. The escape hatch is an asset if one discovers along the way that the compellent attempt was a mistake after all—one misjudged the adversary, or formulated an impossible demand, or failed to communicate what he was doing and what he was after. The escape hatch is an embarrassment, though, if the adversary knows it is there; he can suppose, or hope, that the initiator will turn aside before the risk or pain mounts up.

Still another type is the action that, though beyond recall by the initiator, does not automatically stop upon the victim's compliance. Compliance is a *necessary* condition for stopping the damage but not *sufficient,* and if the damage falls mainly on the adversary, he has to consider what other demands will attach to the same compellent action once he has complied with the initial demands. The initiator may have to promise persuasively that he will stop on compliance, but stoppage is not automatic. Once the missiles are gone from Cuba we may have afterthoughts about antiaircraft batteries and want them removed too before we call off the quarantine or stop the flights.

Finally, there is the action that only the initiator can stop, but can stop any time with or without compliance, a quite "unconnected" action.

In all of these cases the facts may be misperceived by one party or both, with the danger that each may think the other can in fact avert the consequences, or one may fail to do so in the mistaken belief that the other has the last clear chance to avert collision. These different compellent mechanisms, which of course are more blurred and complex in any actual case, usually depend on what the connection is between the threat and the demand—a connection that can be physical, territorial, legal, symbolic, electronic, political, or psychological.

Compellence and Brinkmanship

Another important distinction is between compellent actions that inflict steady pressure over time, with cumulative pain or damage to the adversary (and perhaps to oneself), and actions that impose risk rather than damage. Turning off the water supply at Guantanamo creates a finite rate of privation over time. Buzzing an airplane in the Berlin corridor does no harm unless the planes collide; they probably will not collide but they *may* and if they do the result is sudden, dramatic, irreversible, and grave enough to make even a small probability a serious one.

The creation of risk—usually a shared risk—is the technique of compellence that probably best deserves the name of "brinkmanship." It is a competition in risk-taking. It involves setting afoot an activity that may get out of hand, initiating a process that carries some risk of unintended disaster. The risk is intended, but not the disaster. One cannot initiate *certain* disaster as a profitable way of putting compellent pressure on someone, but one can initiate a moderate *risk* of mutual disaster if the other party's compliance is feasible within a short enough period to keep the cumulative risk within tolerable bounds. "Rocking the boat" is a good example. If I say, "Row, or I'll tip the boat over and drown us both," you'll not believe me. I cannot actually tip the boat over to make you row. But if I start rocking the boat so that it *may* tip over—not because I want it to but because I do not completely control things once I start rocking the boat—you'll be more impressed. I have to be willing to take the risk; then I still have to win the war of nerves, unless I can arrange it so that only you can steady the boat by rowing where I want you to. But it does lend itself to compellence, because one may be able to create a coercive risk of grave consequences where he could not profitably take a deliberate step to bring about those consequences, or even credibly threaten that he would. This phenomenon is the subject of the chapter that follows.

3
THE MANIPULATION
OF RISK

If all threats were fully believable (except for the ones that were completely unbelievable) we might live in a strange world —perhaps a safe one, with many of the marks of a world based on enforceable law. Countries would hasten to set up their threats; and if the violence that would accompany infraction were confidently expected, and sufficiently dreadful to outweigh the fruits of transgression, the world might get frozen into a set of laws enforced by what we could figuratively call the Wrath of God. If we could threaten world inundation for any encroachment on the Berlin corridor, and everyone believed it and understood precisely what crime would bring about the deluge, it might not matter whether the whole thing were arranged by human or supernatural powers. If there were no uncertainty about what would and would not set off the violence, and if everyone could avoid accidentally overstepping the bounds, and if we and the Soviets (and everybody else) could avoid making simultaneous and incompatible threats, every nation would have to live within the rules set up by its adversary. And if all the threats depended on some kind of physical positioning of territorial claims, trip-wires, troop barriers, automatic alarm systems, and other such arrangements, and all were completely infallible and fully credible, we might have something like an old fashioned western land rush, at the end of which—as long as nobody tripped on his neighbor's electric fence and set the whole thing off—the world would be carved up into a tightly bound status quo. The world would be full of literal and figurative frontiers and thresholds that nobody in his right mind would cross.

But uncertainty exists. Not everybody is always in his right mind. Not all the frontiers and thresholds are precisely defined, fully reliable, and known to be so beyond the least temptation to test them out, to explore for loopholes, or to take a chance that they may be disconnected this time. Violence, especially war, is a confused and uncertain activity, highly unpredictable, depending on decisions made by fallible human beings organized into imperfect governments, depending on fallible communications and warning systems and on the untested performance of people and equipment. It is furthermore a hotheaded activity, in which commitments and reputations can develop a momentum of their own.

This last is particularly true, because what one does today in a crisis affects what one can be expected to do tomorrow. A government never knows just how committed it is to action until the occasion when its commitment is challenged. Nations, like people, are continually engaged in demonstrations of resolve, tests of nerve, and explorations for understandings and misunderstandings.

One never quite knows in the course of a diplomatic confrontation how opinion will converge on signs of weakness. One never quite knows what exits will begin to look cowardly to oneself or to the bystanders or to one's adversary. It would be possible to get into a situation in which either side felt that to yield now would create such an asymmetrical situation, would be such a gratuitous act of surrender, that whoever backed down could not persuade anybody that he wouldn't yield again tomorrow and the day after.

This is why there is a genuine risk of major war not from "accidents" in the military machine but through a diplomatic process of commitment that is itself unpredictable. The unpredictability is not due solely to what a destroyer commander might do at midnight when he comes across a Soviet (or American) freighter at sea, but to the psychological process by which particular things become identified with courage or appeasement or how particular things get included in or left out of a diplomatic package. Whether the removal of their missiles from

Cuba while leaving behind 15,000 troops is a "defeat" for the Soviets or a "defeat" for the United States depends more on how it is construed than on the military significance of the troops, and the construction placed on the outcome is not easily foreseeable.

The resulting international relations often have the character of a competition in risk taking, characterized not so much by tests of force as by tests of nerve. Particularly in the relations between major adversaries—between East and West—issues are decided not by who *can* bring the most force to bear in a locality, or on a particular issue, but by who is eventually *willing* to bring more force to bear or able to make it appear that more is forthcoming.

There are few clear choices—since the close of World War II there have been but a few clear choices—between war and peace. The actual decisions to engage in war—whether the Korean War that did occur or a war at Berlin or Quemoy or Lebanon that did not—were decisions to engage in a war of uncertain size, uncertain as to adversary, as to the weapons involved, even as to the issues that might be brought into it and the possible outcomes that might result. They were decisions to embark on a risky engagement, one that could develop a momentum of its own and get out of hand. Whether it is better to be red than dead is hardly worth arguing about; it is not a choice that has arisen for us or has seemed about to arise in the nuclear era. The questions that do arise involve *degrees of risk*—what risk is worth taking, and how to evaluate the risk involved in a course of action. The perils that countries face are not as straightforward as suicide, but more like Russian roulette. The fact of uncertainty—the sheer unpredictability of dangerous events—not only blurs things, it changes their character. It adds an entire dimension to military relations: the manipulation of risk.

There is just no foreseeable route by which the United States and the Soviet Union could become engaged in a major nuclear war. This does not mean that a major nuclear war cannot occur. It only means that if it occurs it will result from a process that is

not entirely foreseen, from reactions that are not fully predictable, from decisions that are not wholly deliberate, from
events that are not fully under control. War has always involved
uncertainty, especially as to its outcome; but with the technology and the geography and the politics of today, it is hard to see
how a major war could get started except in the presence of uncertainty. Some kind of error or inadvertence, some miscalculations of enemy reactions or misreading of enemy intent, some
steps taken without knowledge of steps taken by the other side,
some random event or false alarm, or some decisive action to
hedge against the unforeseeable would have to be involved in
the process on one side or both.[1]

This does not mean that there is nothing the United States
would fight a major war to defend, but that these are things that
the Soviet Union would not fight a major war to obtain. And
there are undoubtedly things the Soviet Union would fight a
major war to defend, but these are not things the United States
would fight a major war to obtain. Both sides may get into a
position in which compromise is impossible, in which the only
visible outcomes would entail a loss to one side or the other so
great that both would choose to fight a major nuclear war. But
neither side wants to get into such a position; and there is nothing presently at issue between East and West that would get
both sides into that position deliberately.

The Cuban crisis illustrates the point. Nearly everybody appeared to feel that there was some danger of a general nuclear
war. Whether the danger was large or small, hardly anyone
seems to have considered it negligible. To my knowledge,
though, no one has ever supposed that the United States or the

---

1. A superb example of this process, one involving local incidents, accidents
of darkness and morning mist, overzealous commanders, troops in panic, erroneous
assessment of damage, public opinion, and possibly a little "catalytic action" by
warmongers, all conjoining to get governments more nearly committed to a war
that might not have been inevitable, occurred within drum-call of my own home.
See the detailed account in Arthur B. Tourtellot, *Lexington and Concord* (New
York, W. W. Norton and Company, 1963). It is chastening to consider that the
"shot heard round the world" may have been fired in the mistaken belief that
a column of smoke meant Concord was on fire.

Soviet Union had any desire to engage in a major war, or that there was anything at issue that, on its merits, could not be settled without general war. If there was danger it seems to have been that each side might have taken a series of steps, actions and reactions and countermeasures, piling up its threats and its commitments, generating a sense of showdown, demonstrating a willingness to carry the thing as far as necessary, until one side or the other began to believe that war had already started, or was so inevitable that it should be started quickly, or that so much was now at stake that general war was preferable to accommodation.

The process would have had to be unforeseeable and unpredictable. If there were some clearly recognizable final critical steps that converted the situation from one in which war was unnecessary to one in which war was inevitable, the step would not have been taken. Alternatives would have been found. Any transition from peace to war would have had to traverse a region of uncertainty—of misunderstandings or miscalculations or misinterpretations, or actions with unforeseen consequences, in which things got out of hand.

There was nothing about the blockade of Cuba by American naval vessels that could have led straightforwardly into general war. Any *foreseeable* course of events would have involved steps that the Soviets or the Americans—realizing that they would lead straightforwardly to general war—would not have taken. But the Soviets could be expected to take steps that, though not leading directly to war, could further compound risk; they might incur some risk of war rather than back down completely. The Cuban crisis was a contest in risk taking, involving steps that would have made no sense if they led predictably and ineluctably to a major war, yet would also have made no sense if they were completely without danger. Neither side needed to believe the other side would deliberately and knowingly take the step that would raise the possibility to a certainty.

What deters such crises and makes them infrequent is that they are genuinely dangerous. Whatever happens to the danger of deliberate premeditated war in such a crisis, the danger of inadvertent war appears to go up. This is why they are called "crises." The essence of the crisis is its unpredictability. The "crisis" that is confidently believed to involve no danger of things getting out of hand is no crisis; no matter how energetic the activity, as long as things are believed safe there is no crisis. And a "crisis" that is known to entail disaster or large losses, or great changes of some sort that are completely foreseeable, is also no crisis; it is over as soon as it begins, there is no suspense. It is the essence of a crisis that the participants are not fully in control of events; they take steps and make decisions that raise or lower the danger, but in a realm of risk and uncertainty.

Deterrence has to be understood in relation to this uncertainty. We often talk as though a "deterrent threat" was a credible threat to launch a disastrous war coolly and deliberately in response to some enemy transgression. People who voice doubts, for example, about American willingness to launch war on the Soviet Union in case of Soviet aggression against some ally, and people who defend American resolve against those doubts, both often tend to argue in terms of a once-for-all decision. The picture is drawn of a Soviet attack, say, on Greece or Turkey or West Germany, and the question is raised, would the United States then launch a retaliatory blow against the Soviet Union? Some answer a disdainful no, some answer a proud yes, but neither seems to be answering the pertinent question. The choice is unlikely to be one between everything and nothing. The question is really: is the United States likely to do something that is fraught with the danger of war, something that could lead—through a compounding of actions and reactions, of calculations and miscalculations, of alarms and false alarms, of commitments and challenges—to a major war?

This is why deterrent threats are often so credible. They do not need to depend on a willingness to commit anything like suicide in the face of a challenge. A response that carries some risk of war can be plausible, even reasonable, at a time when a final, ultimate decision to *have* a general war would be implausible or unreasonable. A country can threaten to stumble into a war even if it cannot credibly threaten to invite one. In fact, though a country may not be able with absolute credibility to threaten general war, it may be equally unable with absolute credibility to forestall a major war. The Russians would have been out of their minds at the time of the Cuban crisis to incur deliberately a major nuclear war with the United States; their missile threats were far from credible, there was nothing that the United States wanted out of the Cuban crisis that the Russians could have rationally denied at the cost of general war. Yet their implicit threat to behave in a way that might—that just might, in spite of all their care and all our care—lead up to the brink and over it in a general war, had some substance. If we were anywhere near the brink of war on that occasion, it was a war that neither side wanted but that both sides might have been unable to forestall.

The idea, expressed by some writers, that such deterrence depends on a "credible first strike capability," and that a country cannot plausibly threaten to engage in a general war over anything but a mortal assault on itself unless it has an appreciable capacity to blunt the other side's attack, seems to depend on the clean-cut notion that war results—or is expected to result—only from a deliberate yes–no decision. But if war tends to result from a *process,* a dynamic process in which both sides get more and more deeply involved, more and more expectant, more and more concerned not to be a slow second in case the war starts, it is not a "credible first strike" that one threatens, but just plain war. The Soviet Union can indeed threaten us with war: they can even threaten us with a war that *we* eventually start, by threatening to get involved with us in a process that blows up into war. And some of the arguments about "superiority" and "inferiority" seem to imply that one of the two sides, being weaker, must absolutely fear war and concede while the other, being stronger, may confidently expect the other to yield. There is undoubtedly a good deal to the notion that the country with the less impressive military capability may be less feared, and the other may run the riskier course in a crisis; other things being equal, one anticipates that the strategically "superior" country has some advantage. But this is a far cry from the notion that the two sides just measure up to each other and one bows before the other's superiority and acknowledges that he was only bluffing. Any situation that scares one side will scare both sides with the danger of a war that neither wants, and both will have to pick their way carefully through the crisis, never quite sure that the other knows how to avoid stumbling over the brink.

Brinkmanship: The Manipulation of Risk

If "brinkmanship" means anything, it means *manipulating the shared risk of war.* It means exploiting the danger that somebody may inadvertently go over the brink, dragging the other with him. If two climbers are tied together, and one wants to intimidate the other by seeming about to fall over the edge, there has to be some uncertainty or anticipated irrationality or it won't work. If the brink is clearly marked and provides a firm footing, no loose pebbles underfoot and no gusts of wind to catch one off guard, if each climber is in full control of himself and never gets dizzy, neither can pose any risk to the other by approaching the brink. There is no danger in approaching it; and while either can deliberately jump off, he cannot credibly pretend that he is about to. Any attempt to intimidate or to deter the other climber depends on the threat of slipping or stumbling. With loose ground, gusty winds, and a propensity toward dizziness, there is some danger when a climber approaches the edge; one can credibly threaten to fall off *accidentally* by standing near the brink.

Without uncertainty, deterrent threats of war would take the form of trip-wires. To incur commitment is to lay a trip-wire, one that is plainly visible, that cannot be stumbled on, and that is manifestly connected up to the machinery of war. And if effective, it works much like a physical barrier. The trip-wire will not be crossed as long as it has not been placed in an intolerable location, and it will not be placed in an intolerable location as long as there is no uncertainty about each other's motives and nothing at issue that is worth a war to both sides. Either side can stick its neck out, confident that the other will not chop it off. As long as the process is a series of discrete steps, taken deliberately, without any uncertainty as to the consequences, this process of military commitment and maneuver would not lead to war. Imminent war—possible war—would be continually threatened, but the threats would work. They would work unless one side were pushed too far; but if the pushing side knows how far that is, it will not push that far.

The resulting world—the world without uncertainty—would discriminate in favor of passivity against initiative. It is easier to *deter* than to *compel*. Among a group of arthritics moving delicately and slowly at a cocktail party, no one can be dislodged from his position near the bar, or ousted from his favorite chair; bodily contact is equally painful to his assailant. By standing in the doorway, one can prevent the entrance or exit of another ailing guest who is unwilling to push his way painfully through.

In fact, without uncertainty all the military threats and maneuvers would be like diplomacy with rigid rules and can be illustrated with a modified game of chess. A chess game can end in win, lose, or draw. Let's change the game by adding a fourth outcome called "disaster." If "disaster" occurs, a heavy fine is levied on both players, so that each is worse off than if he had simply lost the game. And the rules specify what causes disaster: specifically, if either player has moved his knight across the center line and the other player has moved his queen across the center line, the game terminates at once and both players are scored with a disaster. If a white knight is already on the black side of the board when the black queen moves across to the white side, the black queen's move terminates the game in disaster; if the queen was already across when White moved his knight across the center line, the knight's move terminates the game in disaster for both players. And the same applies for the white queen and the black knight.

What does this new rule do to the way a game is played? If a game is played well, and both players play for the best score they can get, we can state two observations. First, a game will never end in disaster. It could only terminate in disaster if one of the players made a deliberate move that he knew would cause disaster, and he would not. Second, the possibility of disaster will be reflected in the players' tactics. White can effectively keep Black's queen on her own side of the board by getting a knight across first; or he can keep both Black's knights on their own side by getting his queen across first. This ability to block or to deter certain moves of the adversary will be an important part of the game; the threat of disaster will be effective, so effective that the disaster never occurs.

In fact, the result is no different from a rule that says no queen can cross a center line if an opponent's knight has already crossed it, and no knight can cross the center line if an opponent's queen has already crossed it. *Prohibitive* penalties imposed on *deliberate* actions are equivalent to ordinary rules.

The characteristic that this chess game shares with the tripwire diplomacy, and that accounts for its peculiar safety, is the absence of uncertainty. There is always some moment, or some final step, in which one side or the other has the last clear chance to turn the course of events away from war (or from disaster in our game of chess) or to turn it away from a political situation that would induce the other to take the final step toward war. The skillful chess player will keep the knight across the center line or near enough to cross before his opponent's queen can get across, with due allowance for the cost of having to devote resources to the purpose. Skillful diplomacy, in the absence of uncertainty, consists in arranging things so that it is one's opponent who is embarrassed by having the "last clear chance" to avert disaster by turning aside or abstaining from what he wanted to do.

But off the chess board the last chance to avert disaster is not always clear. One does not always know what moves of his own would lead to disaster, one cannot always perceive the moves that the other side has already taken or has set afoot, or what interpretation will be put on one's own actions; one does not always understand clearly what situations the other side would not, at some moment, accept in preference to war. When we add uncertainty to this artificial chess game we are not so sure that disaster will be avoided. More important, the risk of disaster becomes a manipulative element in the situation. It can be exploited to intimidate.

To see this, make one more change in the rules. Let us not have disaster occur automatically when queen and knight of opposite color have crossed the center line. Instead, when that occurs, the referee rolls a die. If an ace comes up the game is over and both players are scored with disaster, but if any other number appears the play goes on. If after the next move the queen and knight are still across the center line the dice are rolled again, and so on.

This is a very different game. And not just because disaster may or may not occur when queen and knight get into those positions, instead of occurring with certainty. The difference is that now queen and knight may actually be moved into those positions. One can deliberately move his knight across the line in an attempt to make the queen retreat, if one thinks his adversary is less willing to incur a continuing risk of disaster, or thinks his adversary can be persuaded that oneself will not retreat, and if the momentary risk of disaster is not prohibitive. In fact, getting one's knight across and blocking its return with one's own pieces, so that it clearly takes several moves to retreat, may persuade the adversary that only he, by withdrawing his queen, can reduce the risk within a tolerable time.

If the black queen cannot retreat—if her exit is blocked against timely retreat—the white knight's tactic to force her withdrawal is ineffectual and gratuitously risky. But it can possibly serve another end (another risky one), namely, to enforce "negotiation." By crossing over, once the queen has crossed and cannot readily return, the knight can threaten disaster; White can propose Black's surrender, or a stalemate, or the removal of a bishop or the sacrifice of a pawn. What he gets out of this is wide open; but what began as a chess game has been converted into a bargaining game. Both sides are under similar pressure to settle the game or at least to get the white knight out of mischief. The outcome, it should be noticed, will not necessarily be in White's favor; he created the pressure, but both are subject to the same risk. White's advantage is that he can back out more quickly, as we have set up the game in this example; even he cannot retreat, though, until Black has made his next move, and for the moment both have the same incentive to come to terms. (White's ability to retreat, and Black's inability, may seem more of an advantage to White than it actually is; his ability to retreat is an ability to save both players, equally, from disaster. If no bargain is reached, the white knight has to return, because he is the only one who can. If Black can avoid entering any negotiation—can absent himself from the room or turn off his hearing aid—White's sole remaining objective will be to get his own knight back before he blows things up.) If "disaster" is only somewhat worse, not drastically worse, than losing the chess game, the side that is losing may have more incentive to threaten disaster, or more immunity to the other's threat, and perhaps in consequence a stronger bargaining position. Note, in particular, that all of this has nothing to do with whether a knight is more or less potent than a queen in the chess game; queen and knight can be interchanged in the analysis of this paragraph. If the clash of a squad with a division can lead to unintended war, or of a protest marcher with an armed policeman to an unwanted riot, their potencies are equal in respect of the threats that count.

In this way uncertainty imports tactics of intimidation into the game. One can incur a moderate probability of disaster, sharing it with his adversary, as a deterrent or compellent device, where one could not take, or persuasively threaten to take, a deliberate last clear step into certain disaster.[2]

2. To clarify the theoretical point it may be worth observing that the uncertainty and unpredictability need not arise from a genuine random mechanism like the dice. It is unpredictability, not "chance," that makes the difference; it could as well arise in the clumsiness of the players, some uncertainty about the rules of the game or the scoring system, bad visibility or moves made in secret, the need to commit certain moves invisibly in advance, meddling by a third party, or errors made by the referee. Dice are merely a convenient way to introduce unpredictability into an artificial example.

The route by which major war might actually be reached would have the same kind of unpredictability. Either side can take steps—engaging in a limited war would usually be such a step—that genuinely raise the probability of a blow-up. This would be the case with intrusions, blockades, occupations of third areas, border incidents, enlargement of some small war, or any incident that involves a challenge and entails a response that may in turn have to be risky. Many of these actions and threats designed to pressure and intimidate would be nothing but noise, if it were reliably known that the situation could not get out of hand. They would neither impose risk nor demonstrate willingness to incur risk. And if they definitely would lead to major war, they would not be taken. (If war were desired, it would be started directly.) What makes them significant and usable is that they create a genuine risk—a danger that can be appreciated—that the thing will blow up for reasons not fully under control.[3]

It has often been said, and correctly, that a general nuclear war would not liberate Berlin and that local military action in the neighborhood of Berlin could be overcome by Soviet military forces. But that is not all there is to say. What local military forces can do, even against very superior forces, is to initiate this uncertain process of escalation. One does not have to be able to win a local military engagement to make the threat of it effective. Being able to lose a local war in a dangerous and provocative manner may make the risk—not the sure consequences, but the possibility of this act—outweigh the apparent

3. The purest real-life example I can think of in international affairs is "buzzing" an airplane, as in the Berlin air corridor or when a reconnaissance plane intrudes. The *only* danger is that of an *unintended* collision. The pilot who buzzes obviously wants no collision. (If he did, he could proceed to do it straightforwardly.) The danger is that he may not avoid accident, through mishandling his aircraft, or misjudging distance, or failure to anticipate the movements of his victim. He has to fly close enough, or recklessly enough, to create an appreciated risk that he may—probably won't, but nevertheless may—fail in his mission and actually collide, to everyone's chagrin including his own.

gains to the other side. The white knight is as potent as the black queen in creating a shared risk of disaster.[4]

Limited War as a Generator of Risk

Limited war, as a deterrent to continued aggression or as a compellent means of intimidation, often seems to require interpretation along these lines, as an action that enhances the risk of a greater war. The danger of major war is almost certainly increased by the occurrence of a limited war; it is almost certainly increased by any enlargement in the scope or violence of a limited war that has already taken place. This being so, the threat to engage in limited war has two parts. One is the threat to inflict costs directly on the other side, in casualties, expenditures, loss of territory, loss of face, or anything else. The second is the threat to expose the other party, together with oneself, to a heightened risk of a larger war.

Just how the major war would occur—just where the fault, initiative, or misunderstanding may occur—is not predictable. Whatever it is that makes limited war between great powers a risky thing, the risk is a genuine one that neither side can altogether dispel even if it wants to. To engage in limited war is to start rocking the boat, to set in motion a process that is not al-

4. It may be worth pointing out that, though all attempts to deter or to compel by threat of violence may carry some risk, it is not a necessary character of deterrent threats that they be risky if they are, or try to be, of the full-commitment or tripwire variety discussed in the preceding chapter. What can make them risky is that they may not work as hoped; they are risky because they may fail. Ideally they would carry no risk. It *is* part of the logical structure of the threats discussed in this chapter that they entail risk—the risk of being fulfilled—even though they work (or were about to work) as intended. One is risky the way driving a car is always risky: genuine accidents can always occur, no matter how well the car is designed or how carefully it is driven; risk is a fact of life. The other is risky the way certain forms of roadhogging are risky: a genuine risk is incurred, or created, or enhanced, for the purpose of intimidation, a risk that may not be altogether avoided if intimidation is successfully achieved because it may have to operate for a finite period before compliance brings relief. This risk is part of the *price* of intimidation.

together in one's control. (In the metaphorical language of our chess game, it is to move a queen or a knight across the center line when the other knight or queen is already across, establishing a situation in which factors outside the players' control can determine whether or not the thing blows up.) The risk has to be recognized, because limited war probably does raise the risk of a larger war whether it is intended to or not. It is a consequence of limited war that that risk goes up; since it is a consequence, it can also be a purpose.

If we give this interpretation to limited war, we can give a corresponding interpretation to enlargements, or threats of enlargement, of the war. The threat to introduce new weapons, perhaps nuclear weapons, into a limited war is not, according to this argument, to be judged solely according to the immediate military or political advantage, but also according to the deliberate risk of still larger war that it poses. And we are led in this way to a new interpretation of the trip-wire. The analogy for limited war forces in Europe, or a blockade about Cuba, or troops for the defense of Quemoy, according to this argument, is not a trip-wire that certainly detonates all-out war if it is in working order and fails altogether if it is not. We have something more like a minefield, with explosives hidden at random; a mine may or may not blow up if somebody starts to traverse the field. The critical feature of the analogy, it should be emphasized, is that whether or not one of the mines goes off is at least to some extent outside the control of both parties to the engagement.

This argument is pertinent to the question not only of whether, but of how, to cross the boundaries in some limited war. If one can gently erode a boundary, easing across it with-out creating some new challenge or a dramatic bid for enemy reprisal, and if one finds the current bounds intolerable, that may be the way to do it if one wants the tactical advantages of relaxing a rule. But if the tactical advantages are unimpressive, one's purpose in enlarging some limited war may be to confront the enemy with a heightened risk, to bring into question the possibility of finding new limits once a few have been breached. One may then try not to maximize the stability of new limits as one passes certain thresholds, but to pass them in a way that dramatizes and emphasizes that the engagement is a dangerous one and that the other side should be eager to call a halt. Deliberately raising the risk of all-out war is thus a tactic that may fit the context of limited war, particularly for the side most discontent with the progress of the war. Introduction of nuclear weapons undoubtedly needs to be evaluated in these terms.

Discussions of troop requirements and weaponry for NATO have been much concerned with the battlefield consequences of different troop strengths and nuclear doctrines. But the battlefield criterion is only one criterion, and when nuclear weapons are introduced it is secondary. The idea that European armament should be designed for resisting Soviet invasion, and is to be judged solely by its ability to contain an attack, is based on the notion that limited war is a tactical operation. It is not.

What that notion overlooks is that a main consequence of limited war, and potentially a main purpose for engaging in it, is to raise the risk of larger war. Limited war does this whether it is intended to or not.

This point is fundamental to deterrence of anything other than all-out attack on ourselves. And it is fundamental to the strategy of limited war. The danger of sudden large war—of unpremeditated war—would be a real danger and would obsess the strategic commands on both sides. This danger is enhanced in a crisis, particularly one involving military activity. It is enhanced partly because of the sheer preoccupation with it. And it is enhanced because alarms and incidents will be more frequent, and those who interpret alarms will be readier to act on them.

This is also, to a large extent, the *purpose* of being prepared to fight a local war in Western Europe. The Soviet anticipation of the risks involved in a large-scale attack must include the danger that general war will result. If they underestimate the scale and duration of resistance and do attack, a purpose of resisting is to confront them, day after day, with an appreciation that life is risky, and that pursuit of the original objective is not worth the risk.

This is distantly—but only distantly—related to the notion that we deter an attack limited to Europe by the announced threat of all-out war. It is different because the danger of war does not depend solely on whether the United States would coolly resolve to launch general war in response to a limited attack in Europe. The credibility of a massive American response is often depreciated: even in the event of the threatened loss of Europe the United States would not, it is sometimes said, respond to the fait accompli of a Soviet attack on Europe with anything as "suicidal" as general war. But that is a simple-minded notion of what makes general war credible. What can make it exceedingly credible to the Russians—and perhaps to the Chinese in the Far East—is that the triggering of general war can occur whether we intend it or not.

General war does not depend on our coolly deciding to retaliate punitively for the invasion of Western Europe after careful consideration of the material and spiritual arguments pro and con. General war could result because we or the Soviets launched it in the mistaken belief that it was already on, or in the mistaken or correct belief that, if we did not start it instantly, the other side would. It does not depend on fortitude: it can result from anticipation of the worse consequences of a war that, because of tardiness, the enemy initiates.

And the fear of war that deters the Soviet Union from an attack on Europe includes the fear of a general war that *they* initiate. Even if they were confident that they could act first, they would still have to consider the wisdom of an action that might, through forces substantially outside their control, oblige them to start general war.

If nuclear weapons are introduced, the sensed danger of general war will rise strikingly.Both sides will be conscious of this increased danger. This is partly a matter of sheer expectation; everybody is going to be more tense, and for good reason, once nuclear weapons are introduced. And national leaders will know that they are close to general war if only because nuclear weapons signal and dramatize this very danger—a danger that is self-aggravating in that the more the danger is recognized, the more likely are the decisions that cause war to occur. This argument is neither for nor against the use of nuclear weapons, but for recognizing that this consequence of their use equals in importance—and could far transcend—their tactical battlefield accomplishments.

It is worth noting that this interpretation suggests that the threat of limited war may be potent even when there is little expectation that one could win it.

It is our sheer inability to predict the consequences of our actions and to keep things under control, and the enemy's similar inability, that can intimidate the enemy (and, of course, us too). If we were in complete control of the consequences and knew what would and what would not precipitate war—a war that we started or a war that the enemy started—we could make no threat that did not depend on our ultimate willingness to choose general war.

This is not an argument that "our side" can always win a war of nerves. (The same analysis applies to "their side" too.) It is a reminder that between the alternatives of unsuccessful local resistance on the one extreme, and the fruitless, terrifying, and probably unacceptable and incredible threat of general thermonuclear war on the other, there is a strategy of risky behavior, of deliberately creating a risk that we share with the enemy, a risk that is credible precisely because its consequences are not entirely within our own and the Soviets' control.

Nuclear Weapons and the Enhancement of Risk

The introduction of nuclear weapons raises two issues here. One is the actual danger of general war; the other is the role of this danger in our strategy. On the danger itself, one has to guess how likely it is that a sizable nuclear war in Europe can persist, and for how long, without triggering general war. The danger appears great enough to make it unrealistic to expect a tactical nuclear war to "run its course." Either the nuclear weapons wholly change the bargaining environment, the appreciation of risks, and the immediate objectives, and bring about some termination, truce, tranquilization, withdrawal, or pause; or else the local war very likely becomes swamped in a much bigger war. If these are the likely alternatives, we should not take too seriously a nuclear local war plan that goes to great lengths to carry the thing to its bitter end. There is a high probability that the war either will go down by an order of magnitude or go up by an order of magnitude, rather than run the tactical nuclear course that was planned for it.

More important is how we control, utilize, and react to a sudden increase in the sensed danger of general war. It will be so important to manage this risk properly that the battlefield consequences of nuclear weapons may be of minor importance. The hour-by-hour tactical course of the war may not even be worth the attention of the top strategic leadership.

One can question whether we ought to use nuclear weapons deliberately to raise the risk of general war. But unless we are willing to do this, we should not introduce nuclear weapons against an adversary who has nuclear weapons on his side. This raising of risk is so much of the consequence of nuclear weapons that to focus our planning attention on the battlefield may be to ignore what should be getting our main attention (and what would, in the event, get it). Once nuclear weapons are introduced, it is not the same war any longer. The tactical objectives and considerations that governed the original war are no longer controlling. It is now a war of nuclear bargaining and demonstration.

In a nuclear exchange, even if it nominally involves only the use of "tactical" weapons against tactically important targets, there will be a conscious negotiating process between two very threatening enemies who are worried that the war will get out of hand. The life expectancy of the local war may be so short that neither side is primarily concerned with what happens on the ground within the next day or two. What each side is doing with its strategic forces would be the main preoccupation. It is the strategic forces in the background that provide the risks and the sense of danger; it is they whose disposition will preoccupy national leaders as much as anything that is going on in Europe itself. It is the strategic forces whose minute-by-minute behavior on each side will be the main intelligence preoccupation of the other side.[5]

Limited and localized nuclear war is not, therefore, a "tactical" war. However few the nuclears used, and however selectively they are used, their purpose should not be "tactical" because their consequences will not be tactical. With nuclears, it has become more than ever a war of risks and threats at the highest strategic level. It is a war of nuclear bargaining.

There are some inferences for NATO planning. First, nuclear weapons should not be evaluated mainly in terms of what they could do on the battlefield: the decision to introduce them, the way to use them, the targets to use them on, the scale on which to use them, the timing with which to use them, and the communications to accompany their use should not be determined (or not mainly determined) by how they affect the tactical course of the local war. Much more important is what they do to the expectation of general war, and what rules or patterns of expectations about local use are created. It is much more a war of dares and challenges, of nerve, of threats and brinkmanship, once the nuclear threshold is passed. This is because the danger of general war, and the awareness of that danger, is lifted an order of magnitude by the psychological and military consequences of nuclear explosion.

5. This is why one of the arguments for delegating nuclear authority to theater commanders—as presented in the election campaign of 1964—made little sense. That was the argument that communications between the theater and the American command structure might fail at the moment nuclear weapons were urgently needed. But if the weapons were that urgently needed, especially in the European theater, there would surely be appreciable danger of general war, and to proceed without communicating would guarantee the absence of crucial communication with the Strategic Air Command, the Defense Intelligence Agency, North American Air Defense Command, military forces everywhere, civil defense authorities, and, of course, our diplomatic establishment. It could preclude a choice of what kind of nuclear war to initiate; it could catch the Americans by surprise, and might merely give warning to the Russians.

Second, as a corollary we should not think that the value or likely success of NATO armed forces depends solely, or even mainly, on whether they can win a local war. Particularly if nuclears are introduced, the war may never run its course. Even without the introduction of nuclears, a main function of resistance forces is to create and prolong a genuine sense of danger, of the potentiality of general war. This is not a danger that we create for the Russians and avoid ourselves; it is a danger we share with them. But it is this deterrent and intimidation function that deserves at least as much attention as the tactical military potentialities of the troops.

Third, forces that might seem to be quite "inadequate" by ordinary tactical standards can serve a purpose, particularly if they can threaten to keep the situation in turmoil for some period of time. The important thing is to preclude a quick, clean Soviet victory that quiets things down in short order.

Fourth, the deployment and equipment of nuclear-armed NATO troops, including the questions of which nationalities have nuclear weapons and which services have them, are affected by the purpose and function and character of nuclear and local war. If what is required is a skillful and well-controlled bargaining use of nuclears in the event the decision is taken to go above that threshold, and if the main purpose of nuclears is not to help the troops on the battlefield, it is much less necessary to decentralize nuclear weapons and decisions to local commanders. The strategy will need tight centralized control; it may not require the kind of close battlefield support that is often taken to justify distribution of small nuclears to the troops; and nuclears probably could be reserved to some special nuclear forces.

Fifth, if the main consequence of nuclear weapons, and the purpose of introducing them, is to create and signal a heightened risk of general war, our plans should reflect that purpose. We should plan—in the event of resort to nuclear weapons—for a war of nerve, of demonstration, and of bargaining, not just target destruction for local tactical purposes. Destroying a target may be incidental to the message that the detonation conveys to the Soviet leadership. Targets should be picked with a view to what the Soviet leadership perceives about the character of the war and about our intent, not for tactical importance. A target near or inside the U.S.S.R., for example, is important because it is near or inside the U.S.S.R., not because of its tactical contribution to the European battlefield. A target in a city is important because a city is destroyed, not because it is a local supply or communication center. The difference between one weapon, a dozen, a hundred, or a thousand is not in the number of targets destroyed but in the Soviet (and American) perception of risks, intent, precedent, and implied "proposal" for the conduct or termination of war.

Extra targets destroyed by additional weapons are not a local military "bonus." They are noise that may drown the message. They are a "proposal" that must be responded to. And they are an added catalyst to general war. This is an argument for a selective and threatening use of nuclears rather than large-scale tactical use. (It is an argument for large-scale tactical use only if such use created the level of risk we wish to create.) Success in the use of nuclears will be measured not by the targets destroyed but by how well we manage the level of risk. The Soviets must be persuaded that the war is getting out of hand but is not yet beyond the point of no return.

Sixth, we have to expect the Soviets to pursue their own policy of exploiting the risk of war. We cannot expect the Soviets to acquiesce in our unilateral nuclear demonstration. We have to be prepared to interpret and to respond to a Soviet nuclear "counterproposal." Finding a way to terminate will be as important as choosing how to initiate such an exchange. (We should not take wholly for granted that the initiation would be ours.)

Finally, the emphasis here is that the use of nuclear weapons would create exceptional danger. This is not an argument in favor of their use; it is an argument for recognizing that danger is the central feature of their use.

In other words, nuclears would not only destroy targets but would signal something. Getting the right signal across would be an important part of the policy. This could imply, for example, deliberate and restrained use earlier than might otherwise seem tactically warranted, in order to leave the Soviets under no illusion whether or not the engagement might become nuclear. The only question then would be, how nuclear. It is not necessarily prudent to wait until the last desperate moment in a losing engagement to introduce nuclear weapons as a last resort. By the time they are desperately needed to prevent a debacle, it may be too late to use them carefully, discriminatingly, with a view to the message that is communicated, and with the maintenance of adequate control. Whenever the tactical situation indicates a high likelihood of military necessity for nuclears in the near future, it may be prudent to introduce them deliberately while there is still opportunity to do so with care, selection, and a properly associated diplomacy. Waiting beyond that point may simply increase the likelihood of a tactical use, possibly an indiscriminate use, certainly a decentralized use, determined by the tactical necessities of the battlefield rather than the strategic necessities of deterrence.

In its extreme form the restrained, signaling, intimidating use of nuclears for brinkmanship has sometimes been called the "shot across the bow." There is always a danger—Churchill and others have warned against it—of making a bold demonstration on so small a scale that the contrary of boldness is demonstrated. There is no cheap, safe way of using nuclears that scares the wits out of the Russians without scaring us too. Nevertheless, *any* use of nuclears is going to change the pattern of expectations about the war. It is going to rip a tradition of inhibition on their use. It is going to change everyone's expectations about the future use of nuclears. Even those who have argued that nuclears ought to be considered just a more efficient kind of artillery will surely catch their breath when the first one goes off in anger. Something is destroyed, even if not enemy targets, if ever-so-few nuclears are used. Whatever a few nuclears prove, or fail to prove about their user, they will change the environment of expectations. And it is expectations more than anything else that will determine the outcome of a limited East–West military engagement.

It is sometimes argued, quite correctly, that this tradition can be eroded, and the danger of "first use" reduced, by introducing nuclear weapons in some "safe" fashion, gradually getting the world used to nuclear weapons and dissipating the drama of nuclear explosions. Nuclear depth charges at sea, small nuclear warheads in air-to-air combat, or nuclear demolitions on defended soil may seem comparatively free of the danger of unlimited escalation, cause no more civil disruption than TNT, appear responsible, and set new traditions for actual use, including the tradition that nuclear weapons can be used without signaling all-out war. Obviously to exploit this idea one should not wait until nuclear weapons are desperately needed in a serious crisis, but deliberately initiate them in a carefully controlled fashion at a time and place chosen for the purpose. It might not be wise and might not be practical, but if the intent is to remove the curse from nuclear weapons, this may be the way to do it.

Among the several objections there is one that may be overlooked even by the proponents of nuclear "legitimization." That is the waste involved—the waste of what is potentially the most dramatic military event since Pearl Harbor. President Johnson, remember, referred to a nineteen-year tradition of nonuse; the breaking of that tradition (which grows longer with each passing year) will probably be, especially if it is designed to be, a most stunning event. It will signal a watershed in military history, will instantly contradict war plans and military expectations, will generate suspense and apprehension, and will probably startle even those who make the decision. The first post-Nagasaki detonation in combat will probably be evidence of a complex and anguished decision, an embarkation on a journey into a new era of uncertainty. Even those who propose readier use of nuclear weapons must appreciate that this is so, because of the strong inhibitions they encounter during the dispute.

This is not an event to be squandered on an unworthy military objective. The first nuclear detonation can convey a message of utmost seriousness; it may be a unique means of communication in a moment of unusual gravity. To degrade the signal in advance, to depreciate the currency, to erode gradually a tradition that might someday be shattered with diplomatic effect, to vulgarize weapons that have acquired a transcendent status, and to demote nuclear weapons to the status of merely efficient artillery, may be to waste an enormous asset of last resort. One can probably not, with effect, throw down a gauntlet if he is known to toss his gloves about on every provocation. One may reasonably choose to vulgarize nuclear weapons through a campaign to get people used to them; but to proceed to use them out of expediency, just because they would be tactically advantageous and without regard to whether they ought to be cheapened, would be shortsighted in the extreme.

Face, Nerve, and Expectations

Cold war politics have been likened, by Bertrand Russell and others, to the game of "chicken." This is described as a game in which two teen-age motorists head for each other on a highway —usually late at night, with their gangs and girlfriends looking on—to see which of the two will first swerve aside. The one who does is then called "chicken."

The better analogy is with the less frivolous contest of chicken that is played out regularly on streets and highways by people who want their share of the road, or more than their share, or who want to be first through an intersection or at least not kept waiting indefinitely.

"Chicken" is not just a game played by delinquent teen-agers with their hot-rods in southern California; it is a universal form of adversary engagement. It is played not only in the Berlin air corridor but by Negroes who want to get their children into schools and by whites who want to keep them out; by rivals at a meeting who both raise their voices, each hoping the other will yield the floor to avoid embarrassment; as well as by drivers of both sexes and all ages at all times of day. Children played it before they were old enough to drive and before automobiles were invented. The earliest instance I have come across, in a race with horse-drawn vehicles, antedates the auto by some time:

The road here led through a gully, and in one part the winter flood had broken down part of the road and made a hol-low. Menelaos was driving in the middle of the road, hoping that no one would try to pass too close to his wheel, but Antilochos turned his horses out of the track and followed him a little to one side. This frightened Menelaos, and he shouted at him:

"What reckless driving Antilochos! Hold in your horses. This place is narrow, soon you will have more room to pass. You will foul my car and destroy us both!"

But Antilochos only plied the whip and drove faster than ever, as if he did not hear. They raced about as far as the cast of quoit . . . and then [Menelaos] fell behind: he let the horses go slow himself, for he was afraid that they might all collide in that narrow space and overturn the cars and fall in a struggling heap.

This game of chicken took place outside the gates of Troy three thousand years ago. Antilochos won, though Homer says —somewhat ungenerously—"by trick, not by merit." [6]

Even the game in its stylized teen-age automobile form is worth examining. Most noteworthy is that the game virtually disappears if there is no uncertainty, no unpredictability. If the two cars, instead of driving continuously, took turns advancing exactly fifty feet at a time toward each other, a point would be reached when the next move would surely result in collision. Whichever driver has that final turn will not, and need not, drive deliberately into the other. This is no game of nerve. The lady who pushes her child's stroller across an intersection in front of a car that has already come to a dead stop is in no particular danger as long as she sees the driver watching her: even

6. *The Iliad,* W. H. D. Rouse, transl. (Mentor Books, 1950), p. 273.

if the driver prefers not to give her the right of way she has the winning tactic and gets no score on nerve. The more instructive automobile form of the game is the one people play as they crowd each other on the highway, jockey their way through an intersection, or speed up to signal to a pedestrian that he'd better not cross yet. These are the cases in which, like Antilochos' chariot, things may get out of control; no one can trust with certainty that someone will have the "last clear chance" to avert tragedy and will pull back in time.

These various games of chicken—the genuine ones that involve some real unpredictability—have some characteristics that are worth noting. One is that, unlike those sociable games it takes two to play, with chicken it takes two *not* to play. If you are publicly invited to play chicken and say you would rather not, you have just played.

Second, what is in dispute is usually not the issue of the moment, but everyone's expectations about how a participant will behave in the future. To yield may be to signal that one can be expected to yield; to yield often or continually indicates acknowledgment that that is one's role. To yield repeatedly up to some limit and then to say "enough" may guarantee that the first show of obduracy loses the game for both sides. If you can get a reputation for being reckless, demanding, or unreliable—and apparently hot-rods, taxis, and cars with "driving school" license plates sometimes enjoy this advantage—you may find concessions made to you. (The driver of a wide American car on a narrow European street is at less of a disadvantage than a static calculation would indicate. The smaller cars squeeze over to give him room.) Between these extremes, one can get a reputation for being firm in demanding an appropriate share of the road but not aggressively challenging about the other's half. Unfortunately, in less stylized games than the highway version, it is often hard to know just where the central or fair or expected division should lie, or even whether there should be any recognition of one contestant's claim.[7]

7.   Analytically there appear to be at least three different motivational structures in a contest of "chicken." One is the pure "test case," in which nothing is

Another important characteristic is that, though the two players are cast as adversaries, the game is somewhat collaborative. Even in the stylized version in which they straddle the white line, there is at least an advantage in understanding that, when a player does swerve, he will swerve to the right and not to the left! And the players may try to signal each other to try to coordinate on a tie; if each can swerve a little, indicating that he will swerve a little more if the other does too, and if their speeds are not too great to allow some bargaining, they may manage to turn at approximately the same time, neither being proved chicken.

They may also collaborate in declining to play the game. This is a little harder. When two rivals are coaxed by their friends to have it out in a fight, they may manage to shrug it off skillfully,

---

at stake but reputations, expectations, and precedents. That is, accommodation or obstinacy, boldness or surrender, merely establishes who is an accommodator, who is obstinate or bold, who tends to surrender or what order of precedence is to be observed. A second, not easily distinguished in practice, occurs when something is consciously *put* at stake (as in a gambling game or trial by ordeal) such as leadership, deference, popularity, some agreed tangible prize, or the outcome of certain issues in dispute. (The duel between David and Goliath, mentioned in the note on page 144, is an example of putting something at stake.) The third, which might be called the "real" in contrast to the "conventional," is the case in which yielding or withdrawing yields something that the dispute is about, as in road-hogging or military probes; that is, the gains and losses are part of the immediate structure of the contest, not attached by convention nor resulting entirely from expectations established for future events. The process of putting something at stake—if what is at stake involves third parties—may not be within the control of the participants; nor, in the second and third cases, can future expectations be disassociated (unless, as in momentary road-hogging, the participants are anonymous). So most actual instances are likely to be mixtures. (The same distinctions can be made for tests of *endurance* rather than risk: wealthy San Franciscans were reported to settle disputes by a "duel" that involved throwing gold coins into the bay, one after the other, until one was ready to quit; and the "potlatch" in both its primitive and its contemporary forms is a contest for status and reputation.) A fourth and a fifth case may also deserve recognition: the case of sheer play for excitement, which is probably not confined to teen-agers, and the case of "joint ordeal" in which the contest, though nominally between two (or among more than two) contestants, involves no adversary relation between them, and each undergoes a unilateral test or defends his honor independently of the other's.

but only if neither comes away looking exclusively responsible for turning down the opportunity. Both players can appreciate a rule that forbids play; if the cops break up the game before it starts, so that nobody plays and nobody is proved chicken, many and perhaps all of the players will consider it a great night, especially if their ultimate willingness to play was not doubted.

In fact, one of the great advantages of international law and custom, or an acknowledged code of ethics, is that a country may be obliged *not* to engage in some dangerous rivalry when it would actually prefer not to but might otherwise feel obliged to for the sake of its bargaining reputation. The boy who wears glasses and can't see without them cannot fight if he wants to; but if he wants to avoid the fight it is not so obviously for lack of nerve. (Equally good, if he'd prefer not to fight but might feel obliged to, is to have an adversary who wears glasses. Both can hope that at least one of them is honorably precluded from joining the issue.) One of the values of laws, conventions, or traditions that restrain participation in games of nerve is that they provide a graceful way out. If one's motive for declining is manifestly not lack of nerve, there are no enduring costs in refusing to compete.

Since these tests of nerve involve both antagonism and co-operation, an important question is how these two elements should be emphasized. Should we describe the game as one in which the players are adversaries, with a modest admixture of common interest? Or should we describe the players as partners, with some temptation toward doublecross?

This question arises in real crises, not just games. Is a Berlin crisis—or a Cuban crisis, a Quemoy crisis, a Hungarian crisis, or a crisis in the Gulf of Tonkin—mainly bilateral competition in which each side should be motivated mainly toward winning over the other? Or is it a shared danger—a case of both being pushed to the brink of war—in which statesmanlike forbearance, collaborative withdrawal, and prudent negotiation should dominate?

It is a matter of emphasis, not alternatives, but in distributing emphasis between the antagonistic and the collaborative motives, a distinction should be made. The distinction is between a game of chicken to which one has been deliberately challenged by an adversary, with a view to proving his superior nerve, and a game of chicken that events, or the activities of bystanders, have compelled one into along with one's adversary. If one is repeatedly challenged, or expected to be, by an *opponent* who wishes to impose dominance or to cause one's allies to abandon him in disgust, the choice is between an appreciable loss and a fairly aggressive response. If one is repeatedly forced by *events* into a test of nerve along with an opponent, there is a strong case for developing techniques and understandings for minimizing the mutual risk.

In the live world of international relations it is hard to be sure which kind of crisis it is. The Cuban crisis of October 1962 was about as direct a challenge as one could expect, yet much of the subsequent language of diplomacy and journalism referred to Premier Khrushchev's and President Kennedy's having found themselves together on the brink and in need of statesmanship to withdraw together.[8] The Budapest uprising of 1956 was as near to the opposite pole as one could expect, neither East nor West having deliberately created the situation as a test of nerve, and the Soviet response not appearing as a direct test of Western resolve to intervene. Yet expectations about later American or allied behavior were affected by our declining to acknowledge that events had forced us into a test. This appears to have been a case in which the United States had a good ex-

8. "Brinkmanship" has few friends, "chicken" even fewer, and I can see why most people are uneasy about what, in an earlier book, I called "the threat that leaves something to chance." There is, though, at least one good word to be said for threats that intentionally involve some loss of control or some generation of "crisis." It is that this kind of threat may be more impersonal, more "external" to the participants; the threat becomes part of the environment rather than a test of will between two adversaries. The adversary may find it easier—less costly in prestige or self-respect—to back away from a risky situation, even if we created the situation, than from a threat that is backed exclusively by our resolve and determination. He can even, in backing away, blame us for irresponsibility, or take credit for saving us both from the consequences. Khrushchev was able to claim, after the Cuban crisis, that he had pulled back

cuse to remain outside, and chose even to take that position officially.

The Berlin wall is an ambiguous case. The migration of East Germans can be adduced as the impelling event, not a deliberate Soviet decision to challenge the allied powers. Yet there was something of a dare both in the way it was done and in its being done at all. The Berlin wall illustrates that someone forced into a game of chicken against his better judgment may, if all goes well, profit nevertheless. The U-2 incident of 1960 is interesting in the wealth of interpretations that can be placed on it; a U.S. challenge to Soviet resolve, a Soviet challenge to U.S. resolve, or an autonomous incident creating embarrassment for both sides.

A good illustration of two parties collaborating to avoid being thrust into a test of nerve was the Soviet and American response to the Chinese–Indian crisis of late 1962. It probably helped both sides that they had ready excuses, even good reasons, for keeping their coats on. For anyone who does not want to be obliged into a gratuitous contest, just to preserve his reputation and expectations about future behavior, a good excuse is a great help.

It may seem paradoxical that with today's weapons of speedy destruction brinkmanship would be so common. Engaging in well-isolated small wars or comparatively safe forms of harassment ought to be less unattractive than wrestling on the brink of a big war. But the reason why most contests, military or not, will be contests of nerve is simply that brinkmanship is unavoidable

---

from the brink of war, not that he had backed away from President Kennedy. It is prudent to pull out of a risky situation—especially one that threatens everyone— where it might appear weak to pull away from the threatening opponent. If war could have arisen only out of a deliberate decision by President Kennedy, one based on cool resolve, Khrushchev would have been backing away from a resolved American President; but because the risk seemed inherent in the situation, the element of personal challenge was somewhat diluted. In the same way a rally or a protest march carries the threat of an unintended riot; officials may yield in the interest of law and order, finding it easier to submit to the danger of accident or incident than to submit directly to a threat of deliberate violence.

and potent. It would be hard to design a war, involving the forces of East and West on any scale, in which the risk of its getting out of control were not of commensurate importance with the other costs and dangers involved. Limited war, as remarked earlier, is like fighting in a canoe. A blow hard enough to hurt is in some danger of overturning the canoe. One may stand up to strike a better blow, but if the other yields it may not have been the harder blow that worried him.

How does one get out of playing chicken if he considers it dangerous, degrading, or unprofitable? How would the United States and the Soviet Union, if they both wished to, stop feeling obliged to react to every challenge as if their reputations were continually at stake? How can they stop competing to see who will back down first in a risky encounter?

First, as remarked before, it takes at least two not to play this kind of game. (At least two, because there may be more than two participants and because bystanders have so much influence.) Second, there is no way in the short run that, by turning over a new leaf, one can cease measuring his adversary by how he reacts to danger, or cease signaling to an adversary one's own intentions and values by how one reacts to danger. Confidence has to be developed. Some conventions or traditions must be allowed to grow. Confidence and tradition take time. Stable expectations have to be constructed out of successful experience, not all at once out of intentions.

It would help if each decided not to dare the other again but only to react to challenges. But this will not turn the trick. The definition of who did the challenging will not be the same on both sides. At what point a sequence of actions becomes a deliberate affront is a matter of judgment. Challenges thrust on East and West will never be wholly unambiguous as to whether they were created by one side to test the other or to gain at the other's expense. If all challenges were clear as to origin and could only arise by deliberate intent of the adversary, a conditional cessation would quiet things once for all. But not all crises are so clear in interpretation. And there is too much at stake for either to sit back and be unresponsive for a period long enough to persuade the other that it can safely relax too.

What is at stake is not only the risk of being exploited by one's partner. There is also the risk that the other will genuinely misinterpret how far he is invited to go. If one side yields on a series of issues, when the matters at stake are not critical, it may be difficult to communicate to the other just when a vital issue has been reached. It might be hard to persuade the Soviets, if the United States yielded on Cuba and then on Puerto Rico, that it would go to war over Key West. No service is done to the other side by behaving in a way that undermines its belief in one's ultimate firmness. It may be safer in a long run to hew to the center of the road than to yield six inches on successive nights, if one really intends to stop yielding before he is pushed onto the shoulder. It may save both parties a collision.

It is often argued that "face" is a frivolous asset to preserve, and that it is a sign of immaturity that a government can't swallow its pride and lose face. It is undoubtedly true that false pride often tempts a government's officials to take irrational risks or to do undignified things—to bully some small country that insults them, for example. But there is also the more serious kind of "face," the kind that in modern jargon is known as a country's "image," consisting of other countries' beliefs (their leaders' beliefs, that is) about how the country can be expected to behave. It relates not to a country's "worth" or "status" or even "honor," but to its reputation for action. If the question is raised whether this kind of "face" is worth fighting over, the answer is that this kind of face is one of the few things worth fighting over. Few parts of the world are intrinsically worth the risk of serious war by themselves, especially when taken slice by slice, but defending them or running risks to protect them may preserve one's commitments to action in other parts of the world and at later times. "Face" is merely the interdependence of a country's commitments; it is a country's reputation for action, the expectations other countries have about its behavior. We lost thirty thousand dead in Korea to save face for the United States and the United Nations, not to save South Korea for the South Koreans, and it was undoubtedly worth it. Soviet expectations about the behavior of the United States are one of the most valuable assets we possess in world affairs.

Still, the value of "face" is not absolute. That preserving face—maintaining others' expectations about one's own behavior—can be worth some cost and risk does not mean that in every instance it is worth the cost or risk of that occasion. In particular, "face" should not be allowed to attach itself to an unworthy enterprise if a clash is inevitable. Like any threat, the commitment of face is costly when it fails. Equally important is to help to decouple an adversary's prestige and reputation from a dispute; if we cannot afford to back down we must hope that he can and, if necessary, help him.

It would be foolish, though, to believe that no country has interests in conflict that are worth some risk of war. Some countries' leaders play chicken because they have to, some because of its efficacy. "Nothing ventured, nothing gained." If the main participants wish to stop it, the game can probably be stopped, but not all at once, not without persistence, some luck, and recognition that it will take time. And, of course, there is no guarantee that the cars will not collide.

4

THE IDIOM
OF MILITARY ACTION

Most of the wars we know of have been restrained wars—conditionally restrained, each side's restraint somewhat depending on the enemy's. "Unconditional surrender," the announced aim of the Allies in the Second World War, sounds like an unbounded objective, and the national energies that went into that war were pretty unstinting. But the very idea of "surrender" brings bargaining and accommodation into warfare. Contrast "unconditional surrender" with "unconditional extermination."

Implicit in our demands for surrender was an understanding, a well-grounded expectation, that once Italy, Germany, or Japan laid down its arms it would not be treated to a massacre. In calling that war a restrained one, I have in mind not the unilateral restraint that the Americans or the British showed in Germany or Japan, once they were in charge and the enemy had already surrendered. I have in mind the *conditional* restraint— the bargain, the proposal that we would stop fighting if they would. Italy and Japan, even Germany, could still exact a price in pain and treasure and in postwar stability, and they knew it; they could not win, once the tide had gone against them, but they could make our victory hurt us more or cost us more. The war was costly to both sides, and jointly we could stop it if terms could be negotiated.

Terms could be negotiated; and it has to be remembered that some of the terms were unspoken. The Germans knew that submission meant survival—not slavery, gas chambers, or an endless orgy of plunder and rape by occupation troops, at least not in the Western-occupied portion.

Somebody might argue that Japan was really finished if the

Americans just wanted to wait—that the United States could have gone on producing atomic bombs, dropping them as they became available, and that Japan had nothing to offer in a surrender negotiation. But that would be wrong both in principle and in fact. The United States did want the war finished quickly. The United States did not want to pursue a mass killing in Japan; the Japanese war cabinet would not have been the first government in history to use its own population as a shield, daring an enemy to destroy people as the price of destroying the regime, knowing that violence was done to American principles by obliging an American government virtually to exterminate an enemy. Furthermore, the Japanese had an army in China; their orderly withdrawal depended on organized surrender.

The United States wanted a Japanese government that could order soldiers in the Pacific islands to surrender and not to hold out indefinitely either in continuation of a lost war or as local bandits. The United States wanted the opportunity to impose a stable regime in Japan itself and to conduct a military occupation consistent with its political objectives and democratic principles. The United States wanted a surrender that acknowledged the decisive role of the United States with minimum credit to the Soviet Union and minimum Soviet rights of occupation; that required an early surrender, and one negotiated mainly with the United States. The United States wanted to demobilize a large military establishment and to enjoy the relief that goes with the end of a war; holding an invasion army in readiness for ultimate collapse, while a slow atomic bombing campaign reduced Japan to rubble, was expensive and undesirable. (It was officially believed that invasion would ultimately be necessary unless the Japanese came to terms; whether or not that belief would have held up, it gave powerful reason at the time for strongly preferring an orderly surrender.) The Japanese government, in other words, still had important powers that it could withhold or yield—the capacity to cooperate or not—and therefore had important bargaining assets. The fact that it had infinitely more to lose than did the United States, in case no agreement was reached, should not obscure the fact that the United States could get little consolation out of its ultimate ability to destroy tens of millions of people. In terms of what the United States was bargaining for, the trading position of the Japanese government was not to be despised.

As Kecskemeti points out in his study of the terminal stages of war, "The survival of the loser's authority structure was a necessary condition for the orderly surrender of his remaining forces," creating a dilemma for democratic victors, to whom that authority structure appears the very embodiment of "the enemy."[1] In wars with avowedly limited objectives the preservation of authority on the other side is more readily appreciated. As American troops approached victory on the outskirts of Mexico City in 1847, General Winfield Scott was "persuaded to hold his position and not attempt to force an entry into the City." In his eagerness to secure the fruits of victory, he and his State Department colleague "were easily convinced that a forward movement of the army might cause a general dispersal of officials from the capital, *leaving no one with whom to negotiate*." The fact that they paused too long while the enemy regrouped does not invalidate the principle, but only reminds us that it takes at least as much skill to end a war properly as to begin one to advantage.[2]

The Germans were exhausted, and the French too, when the Franco-Prussian War was brought to a close in 1871. The Germans possessed all that they wanted of French territory and the French had little hope of expelling them; but without complete victory (and often with it) it takes two to stop a war. The French could still exact a price from the Germans, and the Germans from the French. They had a common interest in closing the books on war, cutting their losses or cashing in their gains and putting a stop to the violence. The French wanted the Germans out, and the Germans needed security to evacuate. Both sides had an interest in keeping communications open, respecting emissaries and ambassadors and listening to the other

1. Paul Kecskemeti, *Strategic Surrender* (New York, Atheneum, 1964), p. 24.
2. Otis A. Singletary, *The Mexican War,* pp. 156–57 (italics supplied).

side, and working out reliable arrangements for closing out the war.

Not all of the restraint in these wars was confined to the terminal negotiations. White flags and emissaries have usually been respected, and open cities, ambulances and hospitals, the wounded, the prisoners, and the dead. In battle itself, soldiers have shown a natural willingness to permit, even to encourage, enemy units to come out with their hands up, saving violence on both sides. The character of this restraint, its reciprocal or conditional nature, is even displayed in those instances where it is absent; where no quarter was given, it was usually where none was expected. Even the idea of *reprisal* involves potential restraint—ruptured restraint to be sure, with damages exacted for some violation or excess—but the essence of reprisal is an action that had been withheld, and could continue to be withheld if the other had not violated the bargain.

The striking characteristic of both world wars is that they were unstinting in the use of force; and the restraint—the accommodation, the bargaining, the conditional agreements and the reciprocity—were mainly in the method of termination. The principal boundary to violence was *temporal;* at some point the war was stopped though both sides still had a capacity to inflict pain and cost on the other. Surrender or truce brought the common interest into focus, putting a limit to the losses. But until surrender or truce, the use of force was substantially unbounded.[3]

Contrast the Korean War. It was *fought* with restraint, conscious restraint, and the restraint was on both sides. On the American side the most striking restraints were in territory and weapons. The United States did not bomb across the Yalu (or anywhere else in China) and did not use nuclear weapons. The enemy did not attack American ships at sea (except by shore batteries), bases in Japan, or bomb anything in South Korea, es-

3. The principal exceptions, aside from the treatment of prisoners and other battlefield negotiations, were the reciprocal avoidance of gas, some restraint in the selection of strategic bombing targets early in the war, and the non-exploitation of populations in occupied countries as hostages against invasion.

pecially the vital area of Pusan.[4] And, depending on just who one considers the "enemy" to have been, there were striking inhibitions on nationality. During the first stage of the war, there were no Chinese; and the Soviet Union, with the possible exception of some unacknowledged pilots or technical personnel, never did enter the war with submarines, aircraft, or troops.

The Korean War is our one modern instance of a sizable, overt limited war conducted by well-organized armies representing both sides in the East–West conflict. To call it "restrained" is, of course, to take a very broad view; the density of fire and of manpower was comparable with the campaigns of both world wars. Both sides slugged it out with unrepressed fury: the troops fought for their lives; there was as little etiquette on the battlefield as in any theater of the Second World War; the stakes were high; and there was a strong sense of "showdown." Restraint took the form of specific limitations on the fighting; within those limits, the war was "all out."

It is a strange spectacle, and indeed what makes it plausible is only that it actually occurred. The circumscribed use of force on the Korean peninsula can be understood only by reference to the fearsome threat of violence in the background. Nuclear weapons were known to exist on both sides, East and West, and whatever the estimates about their size and number they scared people; the Soviet Union held in reserve a tidal wave of military manpower and was not believed so vulnerable to attack, even atomic attack, as to be wholly intimidated from launching war in Europe. The consequence was a war in which the fury of battle was exceeded only by the preoccupation with violence held in reserve.

The Korean experience set patterns and precedents that have affected, and will affect, the conduct of limited war and the planning for it. That war not only reflected the phenomenon of

4. According to Halperin, the Chinese were willing to bomb South Korea but only with planes launched from North Korean airfields, and the latter were kept substantially unusable by U.N. air attack. The character of this self-imposed limitation makes the reciprocity especially vivid. Morton H. Halperin, *Limited War in the Nuclear Age* (New York, John Wiley and Sons, 1963), p. 54.

restraint in furious war but undoubtedly determined attitudes toward restraint.

In default of competitors, the Korean War has served as our typical example. It has been distinguished from "all-out war" not only in degree but in kind, at least until Secretary McNamara officially acknowledged that even a major war between the main adversaries could be limited too. Such a prolonged, tightly bounded, energetic, and purely military campaign is at least a possibility in the nuclear era because it actually occurred. But it may be only one possibility, one pattern, one species of a variegated genus of warlike relations, and no more a model of what "limited war" really is than the first animal the Pilgrims saw reflected the wildlife of North America.

Tacit Bargains and Conventional Limits

Nuclear weapons were not used in the Korean War. Gas was not used in the Second World War. Any "understanding" about gas was voluntary and reciprocal—enforceable only by threat of reciprocal use. (That the Geneva Protocol of 1925 outlawed chemical agents in war and was signed by all the European participants in World War II does not itself explain the non-use of gas; it only provided an agreement that both sides could keep if they chose to, under pain of reciprocity.) It is interesting to speculate on whether any alternative agreement concerning poison gas could have been arrived at without formal communication (or even, for that matter, with communication). "Some gas" raises complicated questions of how much, where, under what circumstances; "no gas" is simple and unambiguous. Gas only on military personnel; gas used only by defending forces; gas only when carried by projectile; no gas without warning—a variety of limits is conceivable. Some might have made sense, and many might have been more impartial to the outcome of the war. But there is a simplicity to "no gas" that makes it almost uniquely a focus for agreement when each side can only conjecture at what alternative rules the other side would propose and when failure at coordination on the first try may spoil the chances for acquiescence in any limits at all.

"No nuclears" is simple and unambiguous. "Some nuclears" would be more complicated. Ten nuclears? Why not eleven or twenty or a hundred? Nuclears only on troops in the field? How close to a village can a nuclear be dropped? Nuclears only when the situation is desperate? How desperate is that? Nuclears only on enemy airfields? Why not bridges, too, once the ice is broken? Nuclears only on the Yalu bridge? But once nuclears are available "in principle" for a unique and significant target, won't it be easier to go on and find a second target, and a third, each almost as compelling as the one that preceded it?

There is a simplicity, a kind of virginity, about all-or-none distinctions that differences of degree do not have. It takes more initiative, more soul-searching, more argument, more willingness to break tradition and upset expectations, to do an unprecedented thing once; the second time comes easier, and if the enemy expects you to do it a second time, now that you have done it once, why not do it the second time, and a third?

National boundaries are unique entities. So are rivers. A national boundary marked by a river, as the boundary between Manchuria and North Korea was marked by the Yalu River, is doubly distinctive. It is noticeable and meaningful if somebody conducts military operations, bombing for example, up to the banks of the river; even if he reaches the river only at a few points in his bombing, one is likely to describe the area of military activity as marked or limited by the river. If one looks at a map with pins indicating every place where a bomb has dropped, and tries to see the pattern, he will notice that the pins are all on one side of the river; draw instead an arbitrary irregular line and put all the pins south of it, and the enemy, looking at the pins in his own map on which the line was not drawn, can only be puzzled by the pattern. Bomb once across the Yalu, and the enemy will expect more bombs across the Yalu the next day; keep bombs this side of the Yalu for several months, and the enemy will suppose that, though you may change your mind at any time, the odds are against your bombing north of the Yalu tomorrow.

Even parallels of latitude—arbitrary lines on a map reflecting an ancient number system based on the days in a year, applied o spherical geometry and conventionalized in Western cartography—become boundaries in diplomatic negotiations and conspicuous stopping places in a war. They are merely lines on a map, but they are on *everybody's* map and, if an arbitrary line is needed, lines of latitude are available.

The shoreline is unambiguous. Water is wet and land is dry. Ships come in all sizes and shapes, and so do structures and vehicles on land; but anyone can distinguish the class of ships, which float and are confined to the offshore waters, from the class of objects that rest on hard ground. An artillery piece on the ground may be a fair target, while a gun turret on a floating vessel is "different." It would be hard to draw the line at vessels twenty miles at sea, or over some stipulated tonnage, possibly even between naval vessels and troop transports; but if one draws the line at the shore it is clear what one has done. "No ships" is unambiguous in a way that "some ships" cannot be. In the same way, "no Chinese" is unambiguous in a way that "some Chinese" cannot be. When the Chinese entered the war, they entered in force. The Chinese might have limited their participation to a couple of divisions, at an earlier stage in the war. But who could expect them to stop at two divisions when a third could make a difference? Who would suppose, if the two divisions were identified, that a third might not be lurking somewhere? Who would suppose that, having taken the decision to introduce themselves into the war with two divisions, it required a major new decision to add a third? Who would abide the discovery of two Chinese divisions, still withholding nuclear weapons and keeping this side of the Yalu, yet treat a third Chinese division as an occasion for bombing Manchuria or resorting to nuclear weapons?

And what is so different about nuclear weapons? Is it the size of the explosion? Would everyone expect either side to observe a weight limitation on bombs containing TNT, drawing the line at one ton, or ten tons, or (if there were planes to carry them) fifty tons? And why is a kiloton nuclear bomb so different from an equivalent weight of high explosives dropped in a single attack?

It is. Everybody knows the difference. The difference is not tactical; it is "conventional," traditional, symbolic—a matter of what people will treat as different, of where they will draw the line. There is no physical or military reason to treat a nuclear explosive differently from an explosion of TNT, but there is a symbolic difference that nobody can deny, just as the first mile north of the thirty-eighth parallel is "different" from the last mile south of it. Logistically, an airfield north of the Yalu differs only slightly from an airfield south of the Yalu; and since the planes that operate from it, or the planes that might attack it, do not have to cross a bridge or ride a ferry, the river could just as well be ignored in any tactical analysis. Symbolically, though, there is a gap between them, a difference in kind and not in degree. They belong to different classes of territory, and nobody can ignore the difference. President Johnson said, "Make no mistake. There is no such thing as a conventional nuclear weapon."[5] He was absolutely right; it is by *convention*—by an understanding, a tradition, a consensus, a shared willingness to see them as different—that they are different.

The American participation in the Korean War increased by discrete steps: first, military-aid personnel; then bombing from the air; then a commitment of ground troops. There may be some number of ground troops that is equivalent to a specified air attack, but a commitment of ground troops did not look like just more of the same. Troops were a different class of intervention and signaled more troops on the ground, in a way that intervention by air did not commit us to ground intervention or make it inevitable.

The Yalu was like the Rubicon. To cross it would have signaled something. It was a natural place to stop; crossing it would have been a new start. There are qualitative distinctions between different kinds of military activity, between nuclear weapons and high explosives, between aerial bombing and ground intervention, between ships at sea and installations on shore, between people in the uniform of North Korea and people in the uniform of China. These are discrete, qualitative boundaries, natural lines of demarcation, not necessarily perti-

5. *New York Times,* September 8, 1964, p. 18.

nent in a tactical or a logistic sense, but nevertheless "obvious" places to draw the line, for reasons more related to psychology or custom than to the mathematics of warfare.

What we have is the phenomenon of "thresholds," of finite steps in the enlargement of a war or a change in participation. They are conventional stopping places or dividing lines. They have a legalistic quality, and they depend on precedents or analogy. They have some quality that makes them recognizable, and they are somewhat arbitrary. For the most part they are just "there"; we don't make them or invent them, but only recognize them. These characteristics are not unique to warfare or diplomatic relations. They show up in business competition, racial negotiations, gang warfare, child discipline, and all kinds of negotiated competition. Apparently any kind of restrained conflict needs a distinctive restraint that can be recognized by both sides, conspicuous stopping places, conventions and precedents to indicate what is within bounds and what is out of bounds, ways of distinguishing new initiatives from just more of the same activity.[6] And there are some good reasons why this is so.

The first is that this kind of conflict, whether war or just maneuvering for position, is a process of bargaining—of threats and demands, proposals and counter-proposals, of giving reassurances and making trades or concessions, signaling intent and communicating the limits of one's tolerance, of getting a reputation and giving lessons. And in limited warfare, two things are being bargained over, the outcome of the war, and the mode of conducting the war itself. Just as business firms may "negotiate" an understanding that they will compete by advertising but not by price cuts, and rival candidates may agree implicitly to attack each other's policies but not their private lives; as street gangs may "agree" to fight with fists and stones

6. The phenomenon shows up in the traditional American expression, "the other side of the tracks." Railroad tracks in an American town were not so much a physical barrier to commerce between social classes as a conventional boundary that people could perceive and confidently expect others to perceive. Racial maps of American metropolitan areas show the same striking tendency for black and white to concentrate in areas separated by conspicuous landmarks, usually landmarks whose only significance is that they *are* conspicuous.

but not knives or guns and not to call in outside help; military commanders may agree to accept prisoners of war, and nations may agree to accept limitations on the forces they will commit or the targets they will destroy.

Just as a strike or a price war or a racketeer's stink bomb in a restaurant is *part* of the bargaining and not a separate activity conducted for its own sake, a way of making threats and exerting pressure, so was the war in Korea a "negotiation" over the political status of that country. But, as in most bargaining processes, there was also implicit bargaining about the rules of behavior, about what one would do, or stop doing, according to how the other side behaved.[7]

7.   There seems to be a widespread belief that "negotiation," or "bargaining," is essentially a verbal activity, even a formal one, and that there is no negotiation unless the parties are in direct verbal contact, even face to face.   According to this definition there was no visible "negotiation" between the American government and the Vietcong or the North Vietnamese in the spring of 1965, and no "negotiation" between Khrushchev and Eisenhower in Paris in 1960, in the wake of the U-2 incident, when the summit conference did not quite occur. By the same definition a strike is not part of an industrial negotiation but rather an object of it; to sulk, to walk out, to bang one's shoe on the table, to overturn strikebreakers' automobiles, to concentrate marines in the Caribbean, or to bomb targets in North Vietnam is not only *not* negotiation, according to this definition, but a denial of negotiation— the thing that negotiation is a proper substitute for. For legal or tactical purposes, this is often a good definition; etiquette is worth something, and when the National Labor Relations Act adjures disputants "to bargain in good faith," meaning to sit down and talk responsively, this highly restrictive definition of bargaining serves a purpose, that of imposing some civilized and conservative rules on the conduct of bargaining. Analytically, though, the essence of bargaining is the communication of intent, the perception of intent, the manipulation of expectations about what one will accept or refuse, the issuance of threats, offers, and assurances, the display of resolve and evidence of capabilities, the communication of constraints on what one can do, the search for compromise and jointly desirable exchanges, the creation of sanctions to enforce understandings and agreements, genuine efforts to persuade and inform, and the creation of hostility, friendliness, mutual respect, or rules of etiquette. The actual talk, especially the formal talk, is only a part of this, often a small part, and since talk is cheap it is often deeds and displays that matter most. Wars, strikes, tantrums, and tailgating can be "bargaining" as much as talk can be. Sometimes they are not, when they become disconnected from any conscious process of coercion, persuasion, or communication of intent; but, then, formal diplomatic talk can also cease to be meaningful negotiation.

Much of this bargaining is tacit. Communication is by deed rather than by word, and the understandings are not enforceable except by some threat of reciprocity, retaliation, or the breakdown of all restraint. Because the bargaining tends to be tacit, there is little room for fine print. With ample time and legal resources a line across Korea could be negotiated almost anywhere, in any shape, related or unrelated to the terrain or to the political division of the country or to any conspicuous landmarks. But if the bargaining is largely tacit and there cannot be a long succession of explicit proposals and counterproposals, each side must display its "proposal" in the pattern of its action rather than in detailed verbal statements. The proposals have to be simple; they must form a recognizable pattern; they must rely on conspicuous landmarks; and they must take advantage of whatever distinctions are known to appeal to both sides. National boundaries and rivers, shorelines, the battle line itself, even parallels of latitude, the distinction between air and ground, the distinction between nuclear fission and chemical combustion, the distinction between combat support and economic support, the distinction between combatants and noncombatants, the distinctions among nationalities, tend to have these "obvious" qualities of simplicity, recognizability, and conspicuousness.[8]

Even their arbitrary nature may help. God made the differ-

8. An important aspect of this tacit bargaining is brought out by problems of the following sort. Suppose two persons must agree, without prior communication, on where to draw a line or to impose a limitation. They must do this by proposing, each separately, a line or limitation, and only if they make identical proposals do they succeed in reaching agreement. They look separately at the same map and propose divisions of territory; or they consider various limitations on gas, nuclear weapons, or some other aspect of combat and propose where a line might be drawn. Certain lines or limits prove to be poor candidates: there is no reason for choosing one over the other that is so compelling that one can suppose his partner would make the same choice. Some are good choices—they enjoy uniqueness, or prominence, or some "obvious" quality that makes them stand out as candidates for simultaneous choice. The reader may try it; pick some limitation on, say, nuclear weapons, and let a partner do the same, without any prior understanding but both trying to make the same choice. The results are usually suggestive. To pursue the inquiry further, see Thomas C. Schelling, *The Strategy of Conflict* (Cambridge, Harvard University Press, 1960), pp. 53–80.

ence between land and water; age-old processes of geology made the river; centuries of tradition divided the earth into conventional coordinates of latitude and longitude; man's inability to fly made the difference between air and ground activity a marked one; and in observing these boundaries, one is accepting a kind of outside arbitration, something "natural," something that has the compelling quality of a tradition or precedent and was not just made up for the occasion. A line drawn to the north of the Yalu or to the south of it would have had to be "proposed," while the Yalu itself had only to be "accepted."

Some other qualities are required. The limits must be of a kind that each side can effectively administer on itself; a pilot can recognize a river or a shoreline more easily than some arbitrary line drawn on his map; non-use of a particular weapon, like gas or nuclear arms, is more easily enforced on one's own troops than are particular target limitations that they might ignore or miscalculate in the heat of battle. The limits are most impressive and most likely to stand up if crossing them is conspicuous and would be readily noticed. Most of these considerations reinforce the idea that the particular limitations observed will be qualitative and not matters of degree—distinctive, finite, discrete, simple, natural, and obvious.

Tradition and precedent are important here. (In fact, traditions and precedents themselves have precisely these qualities.) Any particular limitation will be the more expectable, the more recognizable, the more natural and obvious, the more people have got used to recognizing it in the past. The line between nuclears and high explosives was not only *observed* during the Korean War but *reinforced*. The tendency to think of parallels of latitude as obvious "dividing lines" was not only exercised in Korea but reinforced by the experience.

Even the Geneva accords in World War II, governing the treatment of prisoners of war, noncombatants, hospitals, and so forth, though nominally a formal negotiated agreement, have to be recognized as essentially a *tacit* understanding, not in the way the details were worked out but in the way they were accepted and acknowledged during war. A number of countries, including Germany and Britain, had formally subscribed to the code of behavior worked out by the International Committee of the Red Cross; it specified a number of things about treatment of prisoners, how to declare an open city, or how to mark the roof of a hospital. The details of this code had been worked out in advance, with some participation by the countries that ultimately adopted the code. And to a remarkable extent the code was adhered to by countries fighting against each other— remarkable considering that a bitter war was being fought, the conduct of the war was in the hands of "war criminals" in some countries, and the conduct of the war included civilian reprisals and other violent contradictions to the concept of a clean war. If one asks why the Geneva conventions were adhered to, to the extent they were, it is hardly an adequate answer that governments felt morally obliged and politically constrained to be on their good behavior. Moral obligation was notably absent among many participants in the Second World War; and being charged with violation of an "agreement" on the Geneva accords would have been a comparatively minor public relations problem for most of the countries concerned. Evidently there was self-interest in moderating some dimensions of the war, and compliance with the Geneva agreements has to be considered voluntary. It was voluntary and conditional; for the most part countries must have followed the Geneva accords to the extent they did in the interest of reciprocity. But why was it not renegotiated, either by tacit gerrymandering of behavior or by explicit exchange of proposals?

The answer must be that when *some* agreement is needed, and when formal diplomacy has been virtually severed, when neither side trusts the other nor expects agreements to be enforceable, when there is neither time nor place for negotiating new understandings, *any* agreement that is available may have a take-it-or-leave-it quality. It can be accepted tacitly by both sides or by unilateral announcements that one will abide by it if the other does too. Had there been several competing conventions all proposing to govern the treatment of prisoners, each different in detail, it might have been harder to settle on one.

But with a single document available, a single consistent set of procedures already worked out in detail, with no time to renegotiate the fine print, there was a single candidate that could win by default, and otherwise very likely no agreement could be reached.

This interpretation is supported by the fact that the United States complied with the Geneva accords even though it was not a signatory. The United States had no legal obligation to anybody, not having subscribed to that convention. But evidently if the United States wanted to reach an "understanding" with its enemies, at a moment when diplomatic niceties could not be tolerated, the choice was to accept arbitrarily the convention that was available or do without. Typically this is the principal authority behind an arbitrator's suggestion in any dispute: the disputants having reached the point where they cannot satisfactorily negotiate an agreement themselves, and having either called in an arbitrator or had one forced upon them, there is a strong power of suggestion in whatever the arbitrator comes up with. He provides a last chance to settle on the one extant proposal; if agreement is badly desired and further negotiation out of the question, the arbitrator's suggestion may be accepted in default of any alternative.

The sheer arbitrariness undoubtedly helps. The Geneva code already existed, like the thirty-eighth parallel, the Korean shore line, or the Yalu River, and did not need to be proposed, only accepted.

It is worth observing that tacit negotiation of unenforceable agreements can sometimes be much more efficacious than explicit verbal negotiation of agreements that purport to carry some sanction. One difficulty with overt negotiations is that there are too many possibilities to consider, too many places to compromise, too many interests to reconcile, too many ways that the exact choice of language can discriminate between the parties involved, too much freedom of choice. In marriage and real estate it helps to have a "standard-form contract," because it restricts each side's flexibility in negotiation. Tacit bargaining is often similarly restrictive; anything that can't go without saying can't go into the understanding. Only bold outlines can be perceived. Both sides have to identify, separately but simultaneously, a plausible and expectable dividing line or mode of behavior, with few alternatives to choose among and knowing that success on the first try may be essential to any understanding at all. Negotiated truce lines, for example, are rarely as simple as the unnegotiated ones; rivers and coastlines and parallels of latitude, mountain ridges and ancient boundaries are often unambiguous and have power of suggestion behind them. They have to serve if detailed changes cannot be negotiated. This is why some status quo ante is an important benchmark for terminating an affray; benchmarks are needed, and the only ones that will serve are those that both sides can perceive, each knowing that the other perceives them too. In warfare the dialogue between adversaries is often confined to the restrictive language of action and a dictionary of common perceptions and precedents.

The Idiom of Reprisal

Three torpedo boats out of North Vietnamese ports attacked an American destroyer thirty miles off their coast on August 2, 1964. The American ship fought them off, damaging one, and remained in the area. Two days later, accompanied by a sister ship, the destroyer was attacked again, and again the attacking force—this time a larger one—was chased away with the help of American aircraft from a nearby carrier. Twelve hours later, sixty-four American aircraft from the carriers *Ticonderoga* and *Constellation* attacked naval installations in five North Vietnamese ports, reportedly destroying or seriously damaging about half of the fifty PT boats in those harbors and setting fire to a petroleum depot. While the attack was under way, President Johnson announced on television that the North Vietnamese attack had occurred and had to be met with positive reply. "That reply is being given as I speak to you tonight." He said, "Our response for the present will be limited and fitting," adding that, "we seek no wider war" and that he had instructed the Secretary of State to make that position totally clear to friends "and to adversaries."

With only one dissent, the eleven Republican and twenty-two Democratic members of the Senate Foreign Relations and Armed Services Committees were satisfied that the President's decision was "soundly conceived and skillfully executed" and that, in the circumstances, the United States "could not have done less and should not have done more." Republicans and Democrats, military and civilians, even some Europeans, with unusual consensus felt the action was neatly tailored in scope and in character—everybody, that is, with the possible exception of the North Vietnamese and the Communist Chinese. Even they may have thought so. As a matter of fact, theirs was the most important judgment. They were the critics who mattered most. The next step was up to them. America's reputation around the world, both for civilized restraint and for resolve and initiative, was at stake; nevertheless, the most important audience, the one for whose benefit the action was so appropriately designed, was the enemy.

If the American military action was widely judged unusually fitting, this was an almost aesthetic judgment. If words like "repartee" can be applied to war and diplomacy, the military action was an expressive bit of repartee. It took mainly the form of deeds, not words, but the deeds were articulate. The text of President Johnson's address was not nearly as precise and explicit as the selection of targets and the source and timing of the attack. The verbal message reinforced the message delivered by aircraft; and the words were undoubtedly chosen with the Communist as well as the American audience in mind. But that night's diplomacy was carried out principally by pilots, not speechwriters.

War is always a bargaining process, one in which threats and proposals, counterproposals and counterthreats, offers and assurances, concessions and demonstrations, take the form of actions rather than words, or actions accompanied by words. It is in the wars that we have come to call "limited wars" that the bargaining appears most vividly and is conducted most consciously. The critical targets in such a war are in the mind of the enemy as much as on the battlefield; the state of the enemy's expectations is as important as the state of his troops; the threat of violence in reserve is more important than the commitment of force in the field.

Even the outcome is a matter of interpretation. It depends on how the adversaries conduct themselves as much as on the division of spoils; it involves reputations, expectations, precedents broken and precedents established, and whether the action left political issues more unsettled or less unsettled than they were. What happens on the ground (or wherever the limited war takes place) is important, but it may be as important for what it symbolizes as for its intrinsic value. And, like any bargaining process, a restrained war involves some degree of collaboration between adversaries.

No one understands this better than the military themselves, who in many wars have had a high regard for the treatment of prisoners. Aside from decency, there is a good reason for keeping prisoners alive; they can be traded in return for the enemy's captives, or their health and comfort can be made conditional on the enemy's treatment of his own prisoners.[9] Collecting the dead is an even more dramatic instance of the non-conflicting

9. This kind of bargaining power can even be the motive for taking prisoners. At the outbreak of the Peloponnesian War an advance party of Thebans entered Plataea during the night but most were killed or captured; the main force arrived next morning and, since the attack was unexpected, found numbers of Plataeans outside their city. The Thebans planned, therefore, initially to move against the Plataeans outside the walls, "to take some prisoners, so as to have them to exchange, in case any of their own people had been made prisoner." The principle was valid, but the Plataeans beat them to it, announcing, via herald, that *they* already had their prisoners and that, if the Thebans did any harm to the Plataeans outside the walls, they would put to death the Thebans whom they had already taken. They even negotiated a complete Theban withdrawal by offering to return the prisoners, but in the end violated the understanding and killed them. Hannibal tried to sell his prisoners for cash, after the battle of Cannae; the Romans refused, apparently not so much to frustrate Hannibal's economics, though, as to maintain "a precedent indispensable for military discipline." The disciplinary point raised by the Romans is a reminder that bargaining is especially complex when the objects of bargaining are themselves participants, with interests of their own. Livy, *The The War with Hannibal,* pp. 158–65; Thucydides, *The Peloponnesian War,* pp. 97–101.

interest in war, of a recognized common interest among enemies that goes back in history at least to the siege of Troy.[10]

What makes restrained war such a collaborative affair is that so much has to be communicated. The American government not only wanted to conduct the raid on North Vietnamese naval installations; it wanted the North Vietnamese to know why it was doing it and what it was *not* doing. What the North Vietnamese understood from the action—how they interpreted it, what lesson they drew, what they expected next, and what pattern or logic they could see in it—was more important than the dismantling of a minor naval capability. The operation was undoubtedly designed with great care, but it did not take great care to destroy twenty million dollars' worth of North Vietnamese assets and to inflict a few score casualties, or whatever the consequences were. The care was in devising a communication that would be received by the North Vietnamese and the Chinese with high fidelity. And it was in the North Vietnamese interest to read the message correctly: certain things might have

10. Another striking instance is the duel, *as a method of war,* which Yadin finds to have been common in Canaan long before the arrival of the Philistines. "Apparently the stimulus of the duel," he says, "was not primarily boastfulness or conceit on the part of the individual warriors, but the desire of commanders to secure a military decision without the heavy bloodshed of a full-scale battle." He analyzes the familiar example of David and Goliath. "A champion comes forth from the Philistine camp, shouts contemptuously to the Israeli army, and demands that they send a warrior to do battle with him. A close examination of the narrative shows that Goliath is not being simply boastful and provocative. There is a specific intent behind his words. He is offering the army of Israel a method of war which was common enough in his own army but was still strange to the Israeli forces. . . . Goliath is, in fact, suggesting that the contest between him and the representative of Israel shall be *instead* of a battle between the two armies. This finds emphasis in the continuation of his declaration, and he presents the conditions of the contest: 'If he be able to fight with me, and to kill me, then will we be your servants: but if I prevail against him, and kill him, then shall ye be our servants, and serve us.' Here, then," concludes Yadin, "is a form of warfare—a duel—which takes place in accordance with prior agreement of the two armies, both accepting the condition that their fate shall be decided by the outcome of the contest." Goliath's army in the end fled, not keeping the bargain. See Yigael Yadin, *The Art of Warfare in Biblical Lands* (2 vols. New York, McGraw-Hill, 1963), 2, 267–69.

gone wrong with the enterprise, especially in the response to it, that would have been deplorable to *both* sides.

What made the attack on the PT boat appear so "appropriate"? An abstract military evaluation does not tell us much; at a modest cost in casualties (two aircraft lost with their pilots), a modest loss was inflicted on the North Vietnamese military force. Equivalent damage inflicted on the North Vietnamese air force, or its army, or its military supply lines, would not have carried the same meaning and would not have seemed nearly so fitting. What made it seem fitting was not its success as a military threat. It was as an act of reprisal—as a riposte, a warning, a demonstration—that the enterprise appealed so widely as appropriate. Equivalent damage on other military resources might have made as much sense militarily, but the symbolism would have been different.

Had the United States waited a week to mount the attack, some of the connection would have been dissipated. Had the United States struck at North Vietnamese airfields, on grounds that the PT boats had proved ineffectual and the next attack might come from airplanes, the connection would have been less close between act and reprisal. Had the United States returned to the attack day after day, shooting at naval installations, port facilities, and warehouses, the entire operation would have lost neatness; the sensation of "justice" would have been diluted; the "incident" would have been less well-defined; and it would have been harder to tell what was reprisal for the destroyer attack and what was opportunistic military action.

A good way to describe the American response is that it was *unambiguous*. It was articulate. It contained a pattern. If someone asks what the United States did when its destroyers were attacked in the Gulf of Tonkin, there is no disagreement about the answer. One can state the time, the targets, and the weapons used. Nobody supposes that the United States just happened to have an attack on those North Vietnamese ports planned for that day; and nobody is in any doubt about precisely what military action was directly related to the attack on the destroyers.

When a dog on a farm kills a chicken, I understand that the

dead chicken is tied around the dog's neck. If the only purpose for punishing a dog's misdemeanor were to make him suffer discomfort, one could tie a dead chicken around his neck for soiling the rug, or spank him in the living room every time he killed a chicken. But it communicates more to the dog, and possibly appeals to the owner's sense of justice, to make the punishment fit the crime, not only in scope and intensity but in symbols and association.

With the dog we cannot explain; we cannot tie a dead chicken around a dog's neck and *tell* him it is because he bit the postman. We can *tell* the North Vietnamese, though, that we are destroying their PT boats because they attacked our ships, and could just as easily have said, alternatively, that we were shooting up supply routes into Laos, blasting some factories, hitting airbases, or intensifying the war in South Vietnam, in reprisal for their attacks on the destroyers. Why does the *action* have to communicate, as it has to with dogs, when we can perfectly well *verbalize* the connection with every assurance that they are listening?

This is an intriguing question. It seems that governments do feel obliged to make a pattern of their actions, to communicate with the deed as well as with words. In fact, there is probably no characteristic of limited war more striking than this, that one communicates by deed rather than words, or by deed in addition to words, and makes the actions form a pattern of communication in spite of the fact that each side is literate enough to understand what the other is saying. There is something here in the psychology of communication—in people's sense of proportion, of justice, of appropriateness, in the symbolic relation of a response to a provocation, in the pattern that is formed by a coherent set of actions—that goes beyond the abstract military relation between enemies, beyond the economics of cost and damage, beyond the words that are used to rationalize a set of actions. We see it all the time in diplomacy. If the Russians restrict the travel of our diplomats or exclude a cultural visit, our first thought is to tighten restrictions or cancel a visit in return, not to retaliate in fisheries or commerce. There is an *idiom* in this interaction, a tendency to keep things in the same currency, to respond in the same language, to make the punishment fit the character of the crime, to impose a coherent pattern on relations.

It is important to figure out why. It is equally important just to recognize that this is so. Whether it involves power of suggestion, sheer imitation, plain lack of imagination, an irrelevant instinct for coherence and pattern, or instead involves some good reasons—even reasons that are only vaguely understood and responded to instinctively—the phenomenon deserves recognition. When Khrushchev in 1960 complained about U-2 flights, he hinted that Soviet rockets might fire at the bases from which U-2 aircraft were launched in the neighboring countries, Pakistan and Norway. Was this because the U-2 aircraft were a threat to him and he would eliminate it by hitting the bases from which they were flying? Probably not. The analogy of the dead chicken seems at least as compelling as the notion that destroying a particular airfield would in some physical sense prevent a recurrence of the flights. Khrushchev was making a connection, making his threatened riposte peculiarly suit the thing he was complaining against. He could have said he would deny visas to Pakistani, shoot a Pakistani vessel at sea, drop a bomb in a town in Pakistan, sabotage Pakistani railroads, or give military aid to an enemy of Pakistan. But he didn't. He said, or strongly hinted—whether or not he really meant it—that he would blow up the place where the U-2 aircraft took off.

This is so natural, as was the American response to the attack on its destroyers, that we may not even be inclined to question the principle involved. Some of these responses are so "obvious" that one is unaware that "obviousness" constitutes a striking principle of interaction in diplomacy, even of military behavior. It is particularly interesting that this happens between countries that are hostile to each other, between whom the legal niceties need not apply.

To say that it would have been gratuitous, inappropriate, and arbitrary for Khrushchev to destroy an army post after a U-2 flight took off, or for the United States to hit Laotian Communist installations after the destroyers were attacked, only provokes the question: So what? What is the compulsion to embody coherence and pattern in one's actions, especially against somebody who has just tried to shoot up your destroyers or has violated your airspace with a reconnaissance plane? Rules are easy to understand among countries that try to get along with each other, that respect each other, subscribe to a common etiquette and are trying to establish a set of laws to govern their behavior; but when somebody flies U-2 planes over your missile sites, why not kidnap a few of his ballerinas?

Even when this tendency to act in patterns—to respond in the same idiom, to make the punishment fit the crime in character as well as intensity—has been explained it still deserves to be evaluated; the fact that it comes naturally does not mean that it necessarily embodies the highest military or diplomatic wisdom. One could argue, perhaps with some validity, that it results from intellectual laziness: there may be a hundred ways to respond to an enemy action, somehow a choice has to be made, and the choice is easy if the range is narrowed by some tradition or instinct that keeps the game in the same ball park. Rules of etiquette serve this purpose; they limit choice and make life easier. It can also be alleged, possibly with some justification, that bureaucracies have a propensity toward casuistry, legalistic reasoning, and philosophical neatness that makes national leaders instinctively act in a coherent pattern, as though coherence were the same as relevance and as though repartee were the highest form of strategy. The tendency for unfriendly countries to debate with each other continually, to accuse each other and to justify themselves, may enhance this tendency to think of all diplomacy, including military action, as a legalistic adversary proceeding.[11]

11. The urge to make punishment fit the crime in *content* and *nature,* not merely in severity, reportedly appears in children by about age 6 and becomes dominant by age 10 or 12. "The essential point," says Piaget in describing the children's attitudes, "is to do to the transgressor something analogous to what he had done himself, so that he should realize the results of his actions; or again to punish, where it is possible, by the direct material consequences of his misdeed." He calls this the principle

But there is undoubtedly more to it. To relate the reaction to the original action, to impose a pattern on events, probably helps to set limits and bounds. It shows a willingness to accept limits and bounds. It avoids abruptness and novelty of a kind that might startle and excessively confuse an opponent. It maintains a sense of communication, of diplomatic contact, of a desire to be understood rather than misunderstood. It helps an opponent in understanding one's motive, and provides him a basis for judging what to expect as the consequences of his own actions. It helps the opponent to see that bad behavior is punished and good behavior is not, if that is what one wants him to see; unconnected actions, actions chosen at random, might not seem to follow a sequence of cause and effect. In case the opponent might think that one is avoiding the issue, turning aside and pretending not to notice, the direct connection between action and response helps to eliminate the possibility of sheer

---

of "punishment by reciprocity" and identifies it with children's developing notion of social contract, of rules governing relations *among* people rather than rules imposed *on* people by divine or natural authority. The littler children tend to think of rules as something imposed from outside; "expiatory punishment" appeals to them and "there is no relation between the content of the guilty act and the nature of its punishment." On the difference between the two, he says, "The choice of punishments is the first thing that brings this out," the little ones being concerned only with severity, the older ones believing, "not that one must compensate for the offense by a proportionate suffering, but that one must make the offender realize, by means of measures appropriate to the fault itself, in what way he has broken the bond of solidarity."

Of course, "bond of solidarity" somewhat exaggerates the contractual relation between Arabs and Israelis, or Americans and North Vietnamese! Nevertheless, restraints and reprisals among nations are based on little else but reciprocity of some sort; connecting the response to the provocation is a way of showing that, though some "rule" has been violated, rules still exist, are in fact being enforced in the act of reprisal, and are enforceable only by the threat of each other's response. It would be interesting to examine reprisals in colonial areas, where the relation is authoritarian rather than reciprocal, to see whether a more "expiatory" mode of punishment is typical. A key difference between the expiatory and the reciprocal modes is apparently between judging the deed and its punishment as an isolated event and viewing them as episodes in a continuing relationship, an essentially bargaining relationship. See Jean Piaget, *The Moral Judgment of the Child* (New York, Collier Books, 1962), pp. 199–232, especially pp. 206, 217, 227, 232.

coincidence and makes the one appear the consequence of the other.

One can still ask why the same association cannot be made verbally, providing much greater freedom of action if a greater freedom is desired. Part of the answer may be that words are cheap, not inherently credible when they emanate from an adversary, and sometimes too intimate a mode of expression. The action is more impersonal, cannot be "rejected" the way a verbal message can, and does not involve the intimacy of verbal contact. Actions also prove something; significant actions usually incur some cost or risk, and carry some evidence of their own credibility. And actions are less ambiguous as to their origin; verbal messages come from different parts of government, with different nuances, supplemented by "leaks" from various sources and can be contradicted by later verbal messages, while actions tend to be irrevocable, and the fact that action occurred proves that authority is behind it. "I wish it were possible to convince others with words," said President Johnson (April 7, 1965) during the air attacks on North Vietnam, "of what we now find it necessary to say with guns and planes."[12]

The fact that there can be good reason does not imply that this coherent "diplomatic" mode of action is always the best. There can be times when a country wants to shake off the rules, to deny any assurance that its behavior is predictable, to shock the adversary, to catch an adversary off balance, to display unreliability and to dare the opponent to respond in kind, to express hostility and to rupture the sense of diplomatic contact, or even just to have an excuse for embarking on a quite unrelated venture as though it were a rational response to some previous event. This is still diplomacy: there are times to be rude, to break the rules, to do the unexpected, to shock, to dazzle, or to catch off guard, to display offense, whether in business diplomacy, military diplomacy, or other kinds of diplomacy. And

12. "That son of a bitch won't pay any attention to words," is the way President Kennedy said it of Khrushchev. "He has to see you move." Arthur M. Schlesinger, Jr., *A Thousand Days* (Boston, Houghton Mifflin, 1965), p. 391.

there are times when, though in principle one would like to conform to tradition and to avoid the unexpected, the tradition is too restrictive in the choices it offers, and one has to abandon etiquette and tradition, to risk a misunderstanding, and to insist on new rules for the game or even a free-for-all. Even then, the rules and traditions are not irrelevant: breaking the rules is more dramatic, and communicates more about one's intent, precisely because it can be seen as a refusal to abide by rules.

The fact that nations show a tendency to embody their intent in their actions does not mean that this sort of communication is received and interpreted with high fidelity. The Gulf of Tonkin was an extreme case of articulate action, partly because the events were so isolated in space and time from other events and so dramatically at variance with the context. The process of diplomacy by maneuver is typically a good deal clumsier, with actions less subject to careful control for the message they embody, subject to background noise from uncontrollable events, and subject to misinterpretation. Even the Gulf of Tonkin events may not have been as plain to the North Vietnamese at the time as they were shortly afterward to the Armed Services and Foreign Relations Committees.

Tactical Responses and Diplomatic Responses

If one side in a crisis or military engagement steps up the conflict, abandoning some restraint or crossing some threshold, we can distinguish two very different determinants of the other's response. One is the change in the tactical situation— the pressure to avert defeat or to recapture advantage by enlarging its own participation. The other is the incentive to make an overt response, to meet the challenge, to effect a reprisal, to "punish" the other side for its breach of the rules or to "warn" against doing it again, even to force the initiator to back down and observe the old rule, ceasing what he started or withdrawing what he introduced. The Chinese entrance into the Korean War appears mainly motivated by the first determinant, the tactical need to keep the American military from conquering all of Korea. There was no obvious "incident" to which they were reacting, no sudden change in American military conduct that released them from some "obligation" to stay out. In contrast was the American response in the Tonkin Gulf, based not on military requirements but on a diplomatic judgment of what the situation called for. Similarly, when Syrian artillery, which had often harassed Israeli military outposts and received ground fire in return, fired on civilians in late 1964 the Israeli response was to break the ground-fire tradition and use airpower to silence the batteries. I am told that an important consideration in this decision was that a serious departure from routine by the Syrians deserved a retaliatory break in the Israeli tradition—with the attendant risk of enlarging the war.[13]

Because there are these two consequences of any enlargement, the "tactical" effect and the "diplomatic" effect, there are two different modes of enlargement. One is to minimize, the other is to exploit, the appearance of initiative, challenge, rupture, or abandon. If the purpose is to shock the other side, creating a sense of danger, signaling initiative, determination, or even recklessness, and so to intimidate the other and to give it pause, the appearance of "breach in the rules" can be accentuated. Whether the action involves new weapons, nationalities, or targets, shifting from covert to overt intervention, or widening the territorial scope of the war, one mode is to do it suddenly, dramatically, and in a manner that dares the other side to react with equal vigor. Even the choice of what initiative to take can be made with a view to its shock effect rather than its tactical consequences. The use of American "reconnaissance" aircraft in reprisal attacks in Laos in 1964 had mainly this effect; the object was probably not to do a certain amount of damage on the ground but to demonstrate American willingness to become more deeply engaged. Disguising the attacks as Laotian, in a way that permitted the tactical missions to be carried out with equal success, would not have served the purpose.

13. I am also told that the recent American air reprisals in North Vietnam were recognized at the time as a precedent for the interpretation of air reprisals in Syria, air strikes having become incorporated into the language of reprisal with the connotation of once-for-all riposte rather than of a declaration of renewed or enlarged war.

If instead the object is the tactical advantage of relaxing some constraint, not intimidation, an effort can be made to erode or erase the boundary rather than to breach it dramatically. A doctrine of "hot pursuit," for example, can be invoked in attacking enemy airfields or ports, stretching the battlefield in a plausible way. If the enemy then prefers to construe "hot pursuit" not as a gambit in a war of nerves but as a tactical extension of the original war, he may be free to treat it as less in need of reprisal than if the attack were an abrupt change in the character of the conflict.[14]

Manipulation of Conventional Thresholds

Can a government create plausible limitations in advance of a conflict, choosing the kinds it considers safe and advantageous? And can limitations be denied their obvious and compelling quality by being denied in advance or eroded in some fashion? Evidently they can, but not easily.

Take the case of nuclear weapons. Numerous activities fortify the symbolic distinction between nuclear and ordinary explosives. Even the test ban, nominally aimed at peacetime testing, celebrates and ratifies an acknowledged distinction between nuclear weapons and all others. In the same manner, but in reverse, a widespread use of clean nuclear explosives for earth-moving and other economic projects would tend to assimilate nuclear to other explosives, getting people used to the idea that the choice between TNT and nuclear explosives should be based on efficiency alone, eroding the tradition that nuclear explosives are different. Alternatively, the use of nuclear weapons anywhere in combat would shatter a precedent; deliberately introducing them into a war in which they were not necessary, under circumstances where there was little likelihood of things' getting out of hand, could cast doubt on the presumption that nuclear weapons are weapons only of last resort and create ex-

14. For further discussion of "Motives for Expansion and Limitation," see Halperin's chapter with that title, also his chapter on "Interaction Between Adversaries," *Limited War in the Nuclear Age,* pp. 1–25 and 26–38.

pectations that they would be used again when expedient, reducing the plausibility to each nuclear country of the other's reluctance to use them.

Can verbal declarations alone create viable limits? Words apparently do contribute. Directly, the discussion of possible limitations helps to create expectations; it may even do so in the absence of any intent. A sustained argument about whether the use of nuclear weapons at sea, or in air-to-air combat, is qualitatively different from their use on land might create, or sharpen, such a distinction in advance. Verbal activity can call attention to distinctions that might not have been recognized; the United States has usually distinguished between combat assistance and the provision of military advisors, a distinction that might not have been so noticeable without discussion of it. Secretary McNamara's proposal that cities be "off limits" is one that, though its expression does not create the difference between cities and military establishments, can at least call attention to a potential dividing line, announce to the Soviets that this is a distinction the U.S. government is alert to, and propose that information systems be designed to tell the difference between an attack on cities and an attack on military establishments.[15]

Characteristics of Thresholds

Some of these thresholds or limits have the quality that when they are crossed there is unavoidably a dramatic challenge, provoking the question, What will the other do now? Gas and nuclear weapons have this character; one expects the adversary

15. "The open declaration of a city-avoidance option (as compared with mere secret preparation for city avoidance) is, in a sense, a notice served. It is a notice in accordance with which an enemy may well expect the United States to behave in case war is forced upon us. The purpose of making the declaration is not to solicit an 'agreement' to the rules of nuclear war. It is rather to be sure that a potential enemy is aware of the new choice it has. If it values the lives of its citizens, it should take steps to create for itself a targeting option to spare the cities of its enemies." John T. McNaughton, General Counsel of the Department of Defense, address to the International Arms Control Symposium, December 1962, *Journal of Conflict Resolution*, 5 (1963), 233.

not merely to reappraise the tactical situation but to consider some overt reaction, some riposte, some answer. (An obvious possibility is to introduce similar weapons of his own, if he has them.) Another such challenge is crossing a national boundary. Adding forces of a new nationality, especially from a country that is capable of a major increase in military activity (like the Chinese armies in the Korean War) is another dramatic challenge.

An important characteristic of limits or thresholds is whether they apply to both sides. If one breaches a limit (crosses a threshold), is there some equivalent step the other side can take? Is it possible to answer "in kind," or is the particular step unavailable to the other or meaningless for it? In a war in which both the United States and the Soviet Union participated, the introduction, say, of nuclear weapons by one could be "matched" by their use on the other side. ("Matching" would not mean that equivalent nuclear firepower is introduced or that the consequences cancel out, only that there is a response in the same "currency.") In other cases, such as blockade of China, there would be no "obvious" response in kind. Sometimes there are, sometimes not, equivalences or symmetries in the restraints observed and the initiatives available to both sides. If one side cannot simply accept the other's gambit but cannot match it in kind, there may nevertheless be some "obvious" response or reprisal. The introduction of American aircraft carriers into a local war would not be matched by the introduction of carriers by the other side; but attacks *on* the carriers, and on other vessels at sea, might appeal as the "appropriate" response. It is always, of course, up to the governments concerned whether or not to respond in kind; there are the risks of continued escalation, as well as military effectiveness to be taken into account. Nor is there any suggestion that a "matching" response cancels the initial act in a tactical or quantitative sense. Indeed, if there is a strong expectation that the response will be limited to "appropriate" or "obvious" or "matching" actions, the side about to take some new initiative will choose, among the possibilities available, the particular steps in which the other side's "matching response" is comparatively weak—aircraft in Vietnam, for example.

An interesting characteristic of some of these thresholds is that they arise by a historical process, even inadvertently or accidentally, and can acquire status just by coming to be recognized over a prolonged period. The North Koreans, for example, may have considered early in the war that the critical port city of Pusan should be "fair game" but were too busy winning the land war to spare aircraft for the purpose. Modest air attacks on Pusan early in the war might have got the Americans used to them, at a time when they were having to get used to successive defeats on the ground anyhow. Possibly the response would have been reprisals against China, threats of atomic bombs, and so forth; possibly not. But over the weeks this apparent safety of Pusan acquired the status of "sanctuary," and a North Korean attack on the port would have been an abrupt change. In fact, American use of the port—night unloading with lights, for example—reflected an eventual reliance on its sanctuary status, and a bombing attack would have been a breach of expectations.

A more important limitation that acquired status with time was the non-use of nuclear weapons in Korea. In retrospect this was one of major influence: it set a precedent that is fundamental to the inhibition on nuclear weapons today and to the controversies about whether and when nuclear weapons ought to be introduced. Had they been used as a matter of course in Korea—and they might or might not have been decisive, according to how they were used and how the Chinese reacted—there might have been a much greater expectation of nuclear weapons in subsequent engagements, less of a cumulative tradition that nuclear weapons were weapons of last resort.

As a matter of fact, if the United States government had desired to be free to use nuclear weapons whenever it might be expedient—in the straits of Formosa or in Vietnam, in the Middle East or in the Berlin corridor—there would have been a strong case for deliberately using them in Korea even without a military necessity. Their use in Korea could have retarded or eliminated any sensation that nuclear weapons were a different class of weapons; it could have established a precedent that they are to be freely used like any other weapon, would have reduced their revolutionary surprise and shock in subsequent engagements and would have raised the general expectation that, where nuclear weapons would be useful, they would be used. The Korean War itself was decisive in the precedent it set, in its confirmation that the decision to use nuclear weapons was, in a real sense and not just nominally, a matter for presidential decision, and in making nuclear weapons the hallmark of restraint in warfare. In 1964 President Johnson said, "For nineteen peril-filled years no nation has loosed the atom against another. To do so now is a political decision of the highest order."[16] The nineteen years are themselves part of the reason why.

An "Ultimate Limit"?

Some thresholds have made a claim to being the "ultimate limit," the last stopping place before all-out war. There have been several of these, none sacrosanct. The most controversial has been the line between nuclear and non-nuclear weapons; there are many who believe (and official Soviet statements have supported them) that once nuclear weapons are used in any East–West conflict, everything goes. Some think that nuclear weapons would simply be the "signal" that the war was out of hand, that both sides would read the signal and restraint would give way to abandon, each would rush to get in the first strategic blow and the situation would literally explode in all-out war. Others think that nuclear weapons would create so much damage and confusion and so increase the tempo of war that the thing would get out of hand in defiance of the best efforts to control it. Some think that military officers have such an infatuation with firepower and brute force that they would take the bit in their teeth and forget all restraint. And some, without thinking much, just believe that when the first nuclear weapons go off the balloon goes up.

16. *New York Times,* September 8, 1964, p. 18.

Beliefs matter. Beliefs may not correspond to statements; official Soviet declarations that no nuclear war could be limited do not mean that Soviet leaders believe it—or, if they do, that they would not change their minds quickly if a few nuclear weapons went off. President Eisenhower used to say that nuclear weapons ought to be used like artillery, on the basis of efficiency, but that does not begin to imply that he really felt that way; his willingness to negotiate the suspension of nuclear tests is evidence that he was affected by the psychological and symbolic status of nuclear weapons. Even those who believe that the distinction between nuclear and conventional weapons is sentimental folly and political nuisance, and that there is no rational basis for distinguishing explosions according to their internal chemistry, will nevertheless catch their breath when the first one goes off in anger, in a way that cannot be explained merely by the size of the explosion.

But beliefs do matter. If everybody believes, and expects everybody else to believe, that things get more dangerous when the first nuclear weapon goes off, whatever his belief is based on he is going to be reluctant to authorize nuclear weapons, will expect the other side to be reluctant, and in the event nuclear weapons are used will be expectant about rapid escalation in a way that could make escalation more likely. Virtually all of these thresholds are fundamentally matters of beliefs and expectations.

Another "ultimate threshold" that has appealed to some is the direct confrontation of Soviet and American troops in battle. There have been some who felt that a war could be restrained as long as the two major powers were not directly engaged in organized military combat, but that if infantrymen in Soviet and American uniforms, organized in regular units and behaving in accordance with authority, started shooting at each other, that would be "general war" between the United States and the Soviet Union, and it could stop only with the exhaustion or collapse of one side or both in a major war. It is hard to see how even a strong belief in it could have made this true; but it is an interesting bit of testimony to the symbolic character of these thresholds and restraints, and seems not only to make war an extraordinarily "diplomatic" phenomenon but would make the biggest war in the history of mankind a phenomenon of antique diplomacy reminiscent of the dueling etiquette of some centuries ago.[17]

An important "ultimate threshold" that undoubtedly commands more assent is the national boundaries of the United States and the Soviet Union. If "limited war" has meant anything in recent years, it has usually meant a war in which the homelands of the two major adversaries were inviolate. There are undoubtedly many reasons for this, but an important one is surely the principle discussed earlier regarding an attack on California. It is simply the question, "If not here, where?" Is there any stop-

17. As Maxwell Taylor has pointed out, beliefs in certain thresholds can become embodied in the planning process, thereby become reflected in military capabilities and command procedures, and thus become more tangible, more rigid. If a government sufficiently believes that any nuclear war will inevitably become "all out," or that any engagement of Soviet and American troops must become "all out," there may be inadequate plans and inappropriate forces for the contingency that got ignored. In the end it *is* then more dangerous to cross the threshold; choices become more extreme; the threshold is less likely to be crossed but, when crossed, may have to be crossed by a great leap. He strongly implies this was the import of the Soviet-American troop-engagement threshold. As he also implies, there is a tendency for certain terms or concepts to become the subject of "official definitions," and these tend not to be analytical but "legal" in their application.

To say that any Soviet-American engagement in which nuclear weapons are used is "general war," in an official glossary, does not tend to broaden the definition of "general war" to include small engagements, nor does it merely *predict* that small direct nuclear engagements are likely to lead to general war; rather it tends to state that general war *shall* occur under those conditions and that planning to the contrary is unauthorized or contrary to some agreement. This is a fundamental difference between scientific, or analytical, definitions and those that apply to the interpretation of statutes, orders, commitments, and agreements. It is also why any set of "official definitions" is bound to be prejudicial. Maxwell D. Taylor, *The Uncertain Trumpet* (New York, Harper and Brothers, 1960), pp. 7–10, 38–39.

This is not to deny that it *does* make a difference if Americans and Russians find themselves at war or, if not quite "at war," *in* a war on opposite sides. The question of whether or not to bomb North Vietnam's surface-to-air missile sites was recognized to involve, in an attenuated way, the possibility of Russian casualties from American military action;

ping place once the national boundary has been penetrated? Can any limitation of intent be communicated; is there any portion of a country that one can conquer without being tempted to go a little further; is there any portion of one's country that can be yielded without implying that, if a little more pressure is put on, a little more will be yielded?

Visible intent would be important. Suppose Soviet troops spilled into Iran during an uprising in that country, and Turkish or American forces became involved. Soviet aircraft could operate from bases north of the Caucasus, and a possible response would be an attack on those bases by American bombers or possibly missiles. To do more than symbolic damage the missiles would have to contain nuclear warheads; these could be small, detonated at altitudes high enough to avoid fall-out, confined to airfields away from population centers, and might easily make clear to the Soviet government that this was an action limited to the Transcaucasus as an extension of the local theater.

There is no doubt that this would be a risky action. It might or might not be militarily effective; and it might or might not open up some "matching" use of Soviet air strikes, perhaps also with nuclear weapons confined to military targets, possibly including American ships in the Persian Gulf or the Indian Ocean, possibly including Turkish air bases. Even if the war stayed limited it would remain to be analyzed which side, if any, would get a tactical advantage out of the exchange; the question that concerns us here, though, is whether the American air or missile

and while this possibility could have been a significant argument either for or against such bombing, and might not have been decisive either way, it was at least recognized as a significant issue, and properly. It is only the "ultimate" nature of the threshold that is deprecated in the text. The Vietnamese case illustrates that many thresholds can become ambiguous, especially if pains are taken to make them so. Any Russians at the SAM sites were not, presumably, "at war" or even officially "in" it; their presence was more supposed than verified; their participation in the shooting, if any, could be denied by the Soviet Union to reduce the embarrassment to both sides; and in other ways the drama of the "incident" could be played down. The fact that there was no announcement by either side that the sites had been attacked, until several days after the first attack, tended to dilute the incident and make it more casual.

attack on Soviet soil would *necessarily* mean all-out war or anything like it.

It could. It could because the Soviets considered this an intolerable affront and recognized that any failure on their part to respond to the challenge with all-out war would be interpreted as a sign of hopeless weakness, and the United States would become ever more arrogant and intimidating, penetrating Soviet territory whenever it was locally convenient, expecting to use nuclear weapons unilaterally, and knowing that the Soviets having once been pushed to the brink were unwilling to go further. Or it could because the Soviets, obliged to carry out some dramatic reprisal, foresaw a sequence of reprisals spiraling upward with no stopping place; foreseeing that, they might choose to take the initiative. It could also lead to general war because the Soviets responded by reflex, having automatic plans to treat any nuclear attack as all-out war, unable to discriminate a localized attack from a comprehensive one, and thus joining the issue by a process of "automation." Finally, even if the Soviets responded in a moderate way, the thing could still get out of hand. So it could lead to general war, either instantly in a Soviet response or through the further compounding of actions.

But it could also *not* lead to general war. It leads to general war only if the Soviets want it to, or if they believe that it inevitably must. But they needn't believe in the inevitability of it, if only because the American action itself contradicts the premise that both sides consider this to be the signal for general war. The American government surely would not have done it if it intended it to lead to general war; and the American government must not expect the Soviet government to feel that it has to lead to general war. It is an affront, of course, a challenge and a demonstration, and an insult to the integrity of the Soviet Union. But it is also a tactical operation in a local war, one in which Soviet aircraft happen to be operating from bases behind their borders rather than from advanced bases. It is a matter of interpretation whether the Soviets can respond with anything less than general war without seeming to submit; but considering the exorbitant cost of responding with an all-out attack on the

United States, there ought to be strong motives on both sides to interpret the American action as not obliging an all-out response.[18]

Secretary McNamara's policy of 1962 goes even further in suggesting that a major campaign against homelands might still consciously avoid cities. This was a proposal that homelands in the awful emergency of major war not be considered "all-or-none" entities. Even a major attack on military installations need not, according to McNamara's declaration, have to be considered the final, ultimate, step in warfare, bursting the floodgates to an indiscriminate contest in pure destruction. He was talking about a much larger and more violent "limited war" than had theretofore received official discussion, but the principle was the same. What he challenged was the notion that restraint could pertain only to small wars, with a gap or discontinuous jump to the largest of all possible wars, one fought without restraint. His proposal was that restraint could make sense in any war, of any size, and that the traditional distinction between small restrained wars and massive orgies of pure violence, with nothing between, was not logically necessary—was in fact false and dangerous.

Even Secretary McNamara left open, however, the question whether the last frontier in modern warfare, the last threshold before purposeless mutual destruction, occurred at the city limits. One cannot tell from his speech or from his published testimony (or from William Kaufmann's sympathetic exposition of Secretary McNamara's policies)[19] whether or not cities be-

18. There is an important geographical asymmetry between the United States and the Soviet Union; this hypothetical "spillover" of a local war into Soviet territory has few, if any, plausible counterparts in the Western hemisphere. Symbolically the Soviet border might seem almost as dramatic a "last stopping place" as the American shoreline; but tactically and logistically the United States is more remote from most potential theaters of local war, particularly a ground war, and any extension of warfare into United States territory would be a more discontinuous jump. (The most plausible geographical exception might be Florida airbases, had the Cuban crisis become a Caribbean war in 1962.)

19. William W. Kaufmann, *The McNamara Strategy* (New York, Harper and Row, 1964).

came that last category, an all-or-none target list, and whether the very last possibility of reciprocated restraint is to avoid them altogether. Indeed, much published discussion of the "new strategy" that involved possibly sparing cities treated cities the way that many people had treated nuclear weapons, as an ultimate qualitative stopping place beyond which no line could be drawn.

But if it makes sense to take seriously Mr. McNamara's insistence that cities can be distinguished from military installations and a boundary between them possibly observed in war, it equally makes sense to go on and ask whether some deterrence could still work and the war be terminated short of the exhaustion of targets on both sides, even though a city or a few cities had been struck in anger (or by inadvertence, carelessness, inaccuracy, or somebody's failing to cooperate).

Evidently the tempo of nuclear war is what makes people think it hopeless, or unpromising, to keep any relations with the enemy once the first city goes. If cities could be destroyed indefinitely, but at a rate not exceeding one per week or one per day, or even one per hour, nobody could responsibly ignore the possibility that the war might be stopped before both sides ran out of ammunition or cities. An enemy might surrender or come to terms; some truce might be arranged; the original issues that provoked the war might still receive some attention. National leaders could not neglect the fact that millions of people were still alive who would either remain alive or be destroyed according to how negotiations and warfare were conducted. No national leader would think of resigning his job and just turning the dial to "automatic," letting the war run its course, as long as he had a country and a population to which he could feel responsible. And nobody would suppose that the enemy's behavior had been ineluctably fixed in perpetuity by a decision to "go automatic."

But speed cannot be the only reason why cities are widely assumed to go by default, once it appears that the first has gone. Another may be thoughtlessness: if it took years for the government to perceive that even a homeland war, a war fought intercontinentally with nuclear weapons, might be kept within bounds and cities not all deliberately destroyed, maybe it just takes more time for the next question to be asked.

There is a dilemma in dealing with any of these limitations in warfare. Evidently the most powerful limitations, the most appealing ones, the ones most likely to be observable in wartime, are those that have a conspicuousness and simplicity, that are qualitative and not a matter of degree, that provide recognizable boundaries. In fact, a main argument in favor of *any* stopping place is the question mentioned earlier, "If not here, where?" The Americans will not stop at the Yalu, nor the Chinese at the shoreline, nor will any other significant boundaries be recognized and observed if all modes and degrees of participation merge together along an undifferentiated scale. It is undoubtedly in the interest of limiting war that some obvious firebreaks and thresholds occur. To insist that the use of nuclear weapons can as readily be limited at fifty as at zero, or that Hitler could have used just a little gas, or that two Chinese divisions in South Vietnam would have been analyzed for their quantitative significance alone and not with regard to the drama of their nationality, undermines the most important potential limitations. There might be a dramatic threshold separating cities from no cities; and to argue that one can as readily stop after the third city, or the thirteenth, or the thirtieth, detracts from the more promising boundary at zero.

We should remind ourselves that the way this subject is officially discussed will partly determine the answer—whether in fact a war could be contained once a few cities had been destroyed. The critical thing would be whether both sides credit each other with a recognition that the war might still be restrained, or instead credit each other with resignation to the belief that the first city is the final signal of abandon. It could also depend on whether each side had taken the trouble to design its military forces, and its sources of information, so that it could distinguish between a few cities and many and keep its conduct under control. And it would depend on whether, within the category of "cities," everything is but a matter of degree or instead there are some subclasses or patterns or conventional boundaries to help find a stopping place. It *is* hard to stop without an obvious stopping place, and it may be important to search for some; failing any other mode of limitation, something in the pattern and timing of response might help to slow the tempo, to communicate a willingness to bring the war to a close, to maintain a threat in reserve.

The case for a nuclear-conventional distinction is a good deal stronger, though, than the case for drawing a clear line at cities, for the reason that the city limits cannot be all that clear.The chances are good that the distinction between nuclear and conventional weapons will not be blurred, either in prospect or in action. We shall probably know it, and the enemy too, when a nuclear weapon is next used in warfare, whether the initial use is by the United States, by the Soviet Union, or by any other country. But the line separating "cities" from all other localities, and the recognition that a city has in fact been deliberately destroyed, are unlikely to be so clear. (How big a town is a "city"? How near to a city is a military installation "part" of a city? If weapons go astray, how many mistakes that hit cities can be allowed for before concluding that cities are "in" the war? And so on.) It is not a choice between preserving a clean line or blurring it by emphasizing quantitative limitations thereafter; there is no such clean line, and the problem of quantitative limitation, once cities have begun to be hurt, may be like the same problem of stopping at cities in the first place. It means exercising a restraining judgment in the noise and confusion of warfare, not relying on an unambiguous alarm bell that signals a deliberate "jump."

The question was not clearly raised by Secretary McNamara, and to avoid raising it may appear to give an implicit answer in the negative. The answer should not go by default; and while anyone can argue that the first city (if there is a clear "first") raises greatly the likelihood that the rest will turn to cinders (just as one can argue that the first uniformed American soldier shot by a uniformed Russian acting under orders might signal all-out war), this by no means implies that the consequence is so ineluctable that one should take it for granted, nor so nearly inevitable that one should make it genuinely inevitable by treating it as though it were.

Wars of the Battlefield,
Wars of Risk,
and Wars of Pain and Destruction

For a decade American ideas of limited war were dominated by the experience in Korea and by the dangers in Europe. The war in Korea had been mainly a military engagement, not a contest in brinkmanship and not a coercive war of civilian fright and damage. Any war in Europe was expected to be much the same—a military engagement whose outcome would be determined by strength and skill applied on the battlefield, by manpower, firepower, tactical surprise, concentration, and mobility. A European war was not expected to be as protracted as the Korean War; nevertheless, short of blowing up into general war, it was expected to run its course within whatever limits of territory, weaponry, or nationality both sides might set for themselves.

The Cuban crisis raised the prospect of a very different species of "limited war." More than just raising the prospect, probably it should be construed as an actual instance of the new species, one in which no shots were actually fired. This new species is the competition in risk-taking, a military-diplomatic maneuver with or without military engagement but with the outcome determined more by manipulation of risk than by an actual contest of force. The Vietnamese war again brought brinkmanship, this time in the noise of actual warfare rather than the suspense of diplomatic confrontation. The threat of unintended enlargement was evidently meant to intimidate the Chinese and the Russians, to coerce their decisions on whether and how to participate, as well as to pose for the North Vietnamese the risk of a larger war in which destruction or military occupation—even intervention by the Chinese—might cost them much that they had built during the preceding decade, including their own independence. Evidently the risk of some incident involving Russians or Chinese was expected to coerce the United States as well. The risks of such enlargement, with consequences painful to both sides, were so widely remarked that they could hardly have failed to be incorporated in the strategies on both sides.

But the Vietnamese war brought in a new element, new to the United States, if not to Algeria, Palestine, and other arenas outside the East-West competition. This was the direct exercise of the power to hurt, applied as coercive pressure, intended to create for the enemy the prospect of cumulative losses that were more than the local war was worth, more unattractive than concession, compromise, or limited capitulation.

This is yet a third species of limited warfare, and its implications are comparatively unexplored in the strategic literature. One can suppose that, for the same reasons, they may have been comparatively unexplored even within the American government. The Southeast Asian experience will undoubtedly stimulate reflection and analysis of this kind of warfare and possibly lead to controversies and to proposed formulae in the field of punitive warfare somewhat analogous to the familiar controversies and formulae governing limitations on combat, the conventional-nuclear threshold, territories, nationalities, and so forth.

The familiar limits—hypothetical limits for hypothetical wars, or the actual limits in the Korean War—have usually had the static either-or quality discussed earlier; they tend not to be matters of *degree* but matters of type, class, or kind. The classes may be defined in rather arbitrary ways, but they have some appeal to credibility and are pertinent mainly because lines need to be drawn. They can be observed because one side or both may be willing to accept limited defeat or some outcome short of tactical victory, rather than take the initiative in breaching the rules, and prepared to act in a manner that reassures the other side of such willingness. The limits may be respected because, if they are once broken, there is no assurance that any new ones can be found and jointly recognized in time to check the widening of the conflict. They tend to have a legalistic quality in that a particular mode of conduct is perceived to be either inside the limits or outside them, not allowable in some degree and forbidden beyond some limit, not allowable in some degree and progressively less allowable as the quantity or intensity increases.

This raises the important question whether there can be a "degree war" instead of a war limited by types of participants, types of weapons, types of targets, classes of territory, and things of that sort. Might each side watch the intensity of the other's effort, judging it quantitatively, reacting to intensification on the other side with intensification on one's own, adding a few more troops, a few more targets, a few more weapons, not in discrete jumps or by admitting whole classes that correspond to some natural distinctions, but accommodating purely by degree? Or is there a tendency for any mode of warfare, once embarked on, to be intensified up to the next natural limit, up to where a new decision is required to change the quality of participation? What happens to a "limited war" if there are no natural stopping places, no readily perceived limits that both sides can acknowledge, no particular reason for confining the activity at one point rather than another along some quantitative scale?

Reprisal and Hot Pursuit

There are two special cases that fall somewhere on the borderline between qualitative limitations on combat, and the quantitative application of coercive violence. One is reprisal, the other is illustrated by "hot pursuit." The word "reprisal" connotes a response, a reply, a retort, or an answer in kind, and implies a reciprocal action, some punishment for a breach in the rules. The bombing of naval ports in the Gulf of Tonkin had this character; one unaccustomed act was responded to with another, the two actions linked in time with the intended relations of cause and effect, of crime and its punishment, of violation and retribution. Nominally, at least, the reprisal is related to the isolated breach of conduct, not to the underlying continuing dispute. The motivation and intent can of course be more ambitious than that; the object can be a display of determination or impetuosity, not just to dissuade repetition but to communicate a much broader threat. One can even hope for an excuse to conduct the reprisal, as a means of communicating a more pervasive threat. Nevertheless, the reprisal tends to have a direct linkage with a recognizable act, and an association with it in time is intended to communicate that that particular act went beyond certain bounds and has provoked a response that may also go outside the routine bounds, and that the incident is capable of being closed. Reprisals may, of course, spiral in a competition to have the last word, and an exchange of reprisals may become so prolonged as to become disconnected with the original incident and take on a character of coercive warfare, even the character of a showdown. Even then, there may be a tendency for reprisals to retain their linkage in time, each act being adjusted to the one that preceded it. This is different from the steady pressure of coercive warfare, aimed at settling the original dispute and not just at punishing some departure from the accepted method of conducting that dispute. Reprisals often have the function of policing qualitative limits against violation, not of widening the area of combat itself.

A somewhat different function is represented by hot pursuit. In hot pursuit one may chase an intruder back across the line and into his own territory, carrying the fight even to his home base. This, like reprisal, is an isolated event linked to some initiating action of the other side. It is "isolated" in the sense that it does not declare open warfare on the enemy territory so penetrated or on the bases that become momentarily vulnerable to the hot pursuer. The idea of hot pursuit seems to be that it can happen repeatedly, not just once—that is, that the penetration in hot pursuit does not open up a new theater of war. The purpose of verbally invoking "hot pursuit" is not merely to find an excuse for it but to identify a limitation in intent, to let the enemy appreciate that this is not an abandonment altogether of some previous restriction but an allowable departure under the rules of the game.

And it does become part of the rules of the game. This is why it seems to differ from "reprisal." Hot pursuit can become routine; it can become the standard price for becoming engaged with the pursuer. It is not a breach of routine reciprocating an enemy breach, it can be a new routine. It does have the character of a "limitation" defined not by type of action or target or territory but by linkage with an enemy action. It is a qualitative limitation, defined by reference to provocations or opportunities, but with results somewhat like a formula for quantitative limitation.

Hot pursuit is usually thought of in terms of military engagement; reprisals, though the targets may be military, have more the element of punishment and threat. Both differ from the kind of continuous coercive warfare that was introduced by the bombing of North Vietnam in February 1965. That was a bombing campaign, not an isolated event. It was not in response to any particular act of North Vietnam but was an innovation in a war that was already going on, an effort to raise the costs of warfare to North Vietnam and to make them readier to come to terms.

Coercive Warfare

The theoretical question whether limits in warfare tend by nature to be qualitative, or can be matters of degree, assumed sudden relevance with the initiation of the bombing campaign in North Vietnam in February 1965. To approach the question it is helpful to remind ourselves what kind of conflict the Korean War had been. It was almost exclusively a *battlefield* war. There was little coercive exercise of the power to hurt, of the kind discussed in Chapter 1, except to the extent that it hurts when soldiers are killed and wounded and money is spent on war. The civilian pain and destruction were locally devastating, but they were incidental to the battlefield warfare—to the defeat and capture of enemy troops, to the conquest of local territory, to the destruction of supplies and military facilities. Neither side's military strategy was mainly concerned with how much civilian damage was being done locally and whether in view of that damage it made sense to abandon military objectives and to halt the war.

The main power to hurt was held in reserve: the American ability to use nuclear bombs in China, the Soviet ability to hurt the United States or to threaten Western Europe, and any Soviet or Chinese ability to coerce Japan by the threat of bombing, remained latent. The power to hurt surely policed the boundaries of the war, deterring unilateral enlargement and keeping the Soviet Union and the United States from engaging each other directly. But the power to hurt, and susceptibility to hurt in return, circumscribed the war without being deliberately exercised in the conduct of that war.

Contrast the bombing of North Vietnam. This was not an all-out interdiction campaign, exclusively designed to cut supplies to the Vietcong; had it been that, there would have been little reason not to do the bombing on a larger scale at the outset. The bombing had an evident measure of coercive intent behind it: it was evidently designed, at least partly, to inflict plain loss of value on the adversary until he began to behave. The bombing was widely discussed, and sometimes explained by the Administration, as a means of putting pressure on the government of North Vietnam; and when extension to industrial establishments was discussed, it was not mainly in terms of slowing down the enemy's war effort but of raising the cost of not coming to terms. The occasional hints and actual instances of conditional cessation of the bombing testified to its negotiatory character. The results of the bombing in North Vietnam, in contrast to that in the south, were to be sought in North Vietnamese willingness to comply, to accommodate, to withdraw, or to negotiate (as well as in setting a pattern, and possibly a warning, for the contingency of Communist Chinese participation).

The bombing still showed some tendency to stay within class distinctions. (Some of the perceived limits or class distinctions may have arisen by default: if the initial selection of targets omits certain whole areas or types of targets, for reasons of convenience, for momentary lack of a capability to hit them, or for any other reasons incidental to the bargaining process, the mere recognition of these omissions may create a "precedent" that makes dramatic an action that, had it occasionally been engaged in all along, would have received less attention.) The city of Hanoi acquired somewhat the status of a lateral dividing line; bombing north of that city was perceived as a departure from a self-imposed restriction. But once bombs were dropped north of the city there was no sudden concentration there, as though lucrative targets that had been off limits were suddenly declared available. The bombing still took the form of limited coercive punishment, as much in support of negotiation as in support of the military effort, as much a diplomatic move as a military one. There is a suggestion here that coercive warfare can be conducted by degree, in measured doses, in a way that purely military engagements—"battlefield" engagements— tend not to be. When one side can hurt the other, possibly when both can hurt each other, the process may be more gradual, more deliberate, less concentrated.

This may be due to two things. First, the hurting does no good directly; it can work only indirectly. Coercion depends more on the threat of what is yet to come than on damage already done. The pace of diplomacy, not the pace of battle, would govern the action; and while diplomacy may not require that it go slowly, it does require that an impressive unspent capacity for damage be kept in reserve. Unless the object is to shock the enemy into sudden submission, the military action must communicate a continued threat. Furthermore, in a "com-pellent" campaign it may take time for the adversary to comply; decisions depend on political and bureaucratic readjustments; and it may especially take time to arrange a mode of compliance that does not appear too submissive; so diplomacy may dictate a measured pace.

Second, a campaign of civil damage, even when conducted by both sides against each other, will not necessarily be a contest in local strength in which it is important to strike before the other can strike or to concentrate overwhelming force. In military engagements the advantages of surprise, concentration, and timely commitment of reserves usually make it inefficient, perhaps disastrous, to withhold resources too long and to let them dribble slowly into battle. But a campaign of civil damage is often comparatively uncontested, able to be delayed or spread over time with no particular loss in efficiency. Unless there are defenses to be overwhelmed or enemy reinforcements to be preempted, haste may be of no value.

So there may be no great loss in military efficiency, and a gain in diplomatic effectiveness, by limiting the degree of intensity of such a campaign. And if both sides are able to conduct painful reprisals against each other, there may be a natural reluctance to maximize mutual damage.

Most of us, in discussing limited war during the past ten years, have had in mind a war in which *both* sides were somewhat deterred during war itself by unused force and violence on the other side. That is, we were not thinking about wars that were limited because one side was just not interested enough, or one side was so small that an all-out war looked small, or even because one side was restrained or both were by humanitarian considerations. We have mainly been talking about wars that involve some *continued* mutual deterrence, some implicit or explicit understanding about the non-commitment of additional force or non-enlargement to other territories or targets. This mutual deterrence or reciprocity, this conditional withholding or abstention, is important—so important as to deserve the emphasis it got. But in coercive warfare there is another important reason for not committing all of one's force, not destroying all the targets one might destroy, even if one faces no possibility of enemy escalation. It is simply that the object is to make the enemy behave.

To use the threat of further violence against somebody requires that you keep something in reserve—that the enemy still have something to lose. This is why coercive warfare, unless it gets altogether out of hand and becomes vengeful, is likely to look restrained. The object is to exact good behavior or to oblige discontinuance of mischief, not to destroy the subject altogether. This would be true even if the enemy posed no threat of reprisal or reciprocal damage and even if the punitive warfare were costless.

I have been distinguishing the coercive campaign against North Vietnam from the more straightforward military campaign against the Vietcong.The latter campaign will prove to have been substantially "coercive" if in the end the Vietcong yield because their losses are unendurable, to themselves or to their supporters; if in the end they yield or become manageable through sheer loss of numbers, or loss of leadership or supply, it is in my scheme a "battlefield" rather than a coercive mode of warfare that we wage against them. If one insists it is all the same war, it only means the two modes can be mixed in the same war yet be distinguishable and worth distinguishing. South Vietnam has illustrated, as Algeria did dramatically, that one side may wage a coercive and terroristic war while the other tries to oppose it forcibly, not coercively. Algeria showed also that when battlefield warfare proves unavailing, and the terroristic adversary cannot be disarmed, confined, or repelled forcibly, the army that first tries forcible action may resort to terror itself. And Algeria showed that relying on coercive terror in return may prove to be not only degrading but incompatible with the purpose it is intended to serve. North Vietnam suggests the important possibility that coercive warfare may be directed against things the adversary values *other* than his population; when one is trying to coerce governments, rather than populations themselves, the distinction between civilian (non-military) targets and civilians themselves is a crucial one.

Coercive Warfare and Compellence

Among the reasons why coercive warfare has not figured much in our theoretical discussions or our military plans, one is that we have been mainly concerned with "deterrence," and deterrence is comparatively simple. Partly our aim has indeed been deterrence; partly deterrence has been a euphemism for the broader concept of coercion, as "defense" has replaced words like "war" and "military" in our official terminology. It is a restrictive euphemism if it keeps us from recognizing that there is a real difference between deterrence and what, in Chapter 2, I had to call "compellence," that is, a real difference between inducing inaction and making somebody perform.

Compellence is the business the United States got into in North Vietnam. It was trying to make the North Vietnamese regime do something (even if only to stop something it was doing) and that is different from deterrence."Compellence" helps to explain why the coercion took the form of delivered damage, not just verbal threats of damage; the Americans communicated the threat by progressive fulfillment, because the first step was up to the Americans. Compellence also helps to explain why this kind of campaign needs to be allocated over time and apportioned in its intensity, in a way that retaliatory deterrent threats often do not. And it explains why it is so important to know who is in charge on the other side, what he treasures, what he can do for us and how long it will take him, and why we have the hard choice between being clear so that he knows what we want or vague so that he does not seem too submissive when he complies. Compellence was new business for the United States in Vietnam, as was coercive warfare; it was no coincidence that these two departures from earlier concepts were linked together in place and time.

This was a new departure undertaken in rather specialized circumstances. First, the bombing itself was unilateral; the North Vietnamese were militarily unable to do anything like responding in kind. How such a war might go if both sides were capable of conducting similar and simultaneous campaigns against each other received no answer. Second, the Vietcong had already been using terroristic techniques of intimidation, against civilians as well as against enemy military personnel, and the war had never been confined to straightforward engagement. Third, nuclear weapons were not used; the weapons most peculiarly suited to civil destruction and the ones whose reciprocated use could accelerate most rapidly and get out of hand were not involved.

In fact, there was no hint that nuclear weapons were being considered in this role. But of course they would have to be considered if the adversary were China rather than North Vietnam, and undoubtedly would be considered, both for their greater effectiveness against a larger adversary and because it would become a much more serious war.

Deterrence will go on being our main business, compellence the exception; but for actual warfare the historical mixture may now be more nearly representative of what we have to expect. Roughly speaking we have one limited war of the battlefield (Korea), we have several contests in risk-taking (Berlin, Cuba), and we have one example of coercive violence, North Vietnam. We'd be wise to recognize that North Vietnam may be as "typical" of limited war as Korea, and then turn around and look at other parts of the world in the light of this new emphasis. I see no reason to suppose that a war in Europe, if it should break out, would be a battlefield test of strength the way Korea was rather than a competition in risk taking, as Cuba was, or a coercive campaign, as North Vietnam has been. In a way, because of the greater relevance of nuclear weapons, one might put greater emphasis on brinkmanship and coercive civil damage than on battlefield tactics, when thinking about Europe. The Vietnamese war provided a precedent for taking this seriously.

Coercive Nuclear Warfare

The relevance to Europe of coercive warfare, war involving limited coercive campaigns of civil destruction and not just qualitative limits on the modes of combat, needs to be examined. And there the widespread expectation that nuclear weapons would be used in a serious war may change the picture substantially. If strategies for the conduct of limited war in Europe have been dominated by the Korean experience, and the possibility of coercive warfare or a war of reprisal excluded by default, there should be an examination of whether this may be the kind of warfare that in fact would come about. The Vietnamese war does suggest that if the war is going badly for a participant who has the option of shifting the emphasis from combat to coercion, he may do so; doing so with nuclear weapons might be ever so much more dangerous, but might appear to the loser correspondingly more potent. In thinking about a nuclear strategy for Europe we ought, then, to consider whether a nuclear war might degenerate (or transcend) into a coercive campaign rather than retain the characteristics of a test of strength on the battlefield.

Because nuclear weapons are peculiarly suited to the creation of pain and damage and fright there is some presumption that if they were used they would be used, wittingly or unwittingly, to hurt, to intimidate, to coerce.

The suggestion has occasionally been made that if the Chinese or North Koreans again attack South Korea, if the Soviet Union attacks Western Europe or Iran, if the Chinese attack India, it may not be necessary to oppose them with force. A little violence may do the trick. Knock out a city, tell them to quit; knock out another if they don't, and keep it up until they do. The earliest proposal I know of, and a provocative one, was by Leo Szilard, who delighted in putting his ideas in shockingly pure form. As early as 1955 he proposed that if the Soviets invaded a country that we were committed to protect, we should destroy a Soviet city of appropriate size. In fact he even suggested that we publish a "price list" indicating to the Soviets what it would cost them, in population destroyed, to attack any country on the list. On whether the Soviets might be motivated to destroy a city of ours in return, Szilard allowed that they probably would be; that was part of the price. They would get little consolation from it, he argued, and our willingness to lose a city in return would be testimony of our resolve. A cold-blooded willingness to punish the enemy for his transgressions, even if it hurt us as much as them, he considered an impressive display.[20]

In less artificial form the notion of "limited reprisal" or "limited retaliation," a "limited punitive war" or "limited strategic war," has been broached from time to time by theorists but never discussed, as far as I know, by officials in any country (although Khrushchev, as mentioned above, during the U-2 dispute of 1960 hinted that he might punitively fire rockets at

20. Leo Szilard, "Disarmament and the Problem of Peace," *Bulletin of the Atomic Scientists, 11* (1955), 297–307.

the bases from which the U-2 flights were launched). Not only has there been official silence on it, but the possibility of using pure violence rather than fighting a local and limited "all-out" military war confined in territory, weapons, and nationalities, has received little unofficial attention. The idea stayed alive in footnotes and occasional tentative articles, was momentarily dignified by nine authors who produced a book on it in 1962, [21] still received little notice and has never become one of the standard categories in the analysis of warfare.

But if we can talk about wars in which tens of millions could be killed thoughtlessly, we ought to be able to talk about wars in which hundreds of thousands might be killed thoughtfully. A war of limited civilian reprisal can hardly be called "unrealistic"; there is no convincing historical evidence that any particular kind of nuclear warfare is realistic. What often passes for realism is conversational familiarity. Any kind of war that is discussed enough becomes familiar, seems realistic, and is granted some degree of likelihood; types of war that have not been discussed have a novelty that makes them "unrealistic." Of course, if a style of warfare has not been thought about, it may never occur—unless it is the kind that becomes suddenly plausible in a crisis, or can be eased into without deliberate intent.

The idea, though, that war can take the form of measured punitive forays into the enemy's homeland, aimed at civil damage, fright, and confusion rather than tactical military objectives, is not new; it may be the oldest form of warfare. It was standard practice in Caesar's time; to subdue the Menapii, a troublesome tribe in the far north of Gaul, he sent three columns into their territory, "burning farms and villages, and taking a large number of cattle and prisoners. By this means the Menapii were compelled to send envoys to sue for peace." [22]

---

21. Klaus Knorr and Thornton Read, ed., *Limited Strategic War* (Princeton, Princeton University Press, 1962).

22. *The Conquest of Gaul,* pp. 164–65. See also pp. 115–18 on "the first crossing of the Rhine," where "his strongest motive was to make the Germans less inclined to come over into Gaul by giving them reason to be alarmed on their own account, and showing them that Roman armies could and would advance across the river."

Nor were punitive reprisals confined to relations between a colonial power and its subjects; Oman describes this form of warfare between the Byzantines and the Saracens in the ninth century. When the Saracen invaded,

> much could also be done by delivering a vigorous raid into his country and wasting Cilicia and northern Syria the moment his armies were reported to have passed north into Cappadocia. This destructive practice was very frequently adopted, and the sight of two armies each ravaging the other's territory without attempting to defend its own was only too familiar to the inhabitants of the borderlands of Christendom and Islam.[23]

Coercive warfare of this sort not only characterized the struggle in Algeria, and the Arab–Israeli cold-war relationship; it has been present in greater or lesser degree in strategies of intimidation ranging from lynching to strategic bombing.

Actual violence is rarely as pure in character or purpose as a

---

23. Oman, *The Art of War in the Middle Ages,* p. 42. It is important to distinguish two variants of this strategy. One is to coerce the enemy's behavior directly by the threat of damage—to make him quit or surrender, in much the same way that hostages are used. The other is to oblige him to bring his offensive forces home (or to keep them home) in a defensive role and to give up or curtail his original campaign. This latter purpose seems to have been among Caesar's motives for his campaign across the Rhine (p. 115); and the same principle when inverted—forcing troops to leave the security of their town walls and sally forth to battle—was among the motives for crop destruction and other marauding activities practiced by invaders in ancient times. (There is an anecdote of a general, on the defensive, being taunted by the opposing general, "If you are a great general, come down and fight it out." His reply was, "And if you are a great general, make me fight it out against my will.") The tactic of forcing, by attacks on life and property, an enemy to commit himself to battle is often crucial to guerilla operations and—when it can be done—to counterguerilla operations. And it was a major consequence (although not recognized at the time) of the strategic bombing of Germany. Aside from the damage inflicted, the bombing raids caused the Germans to divert, by the time of the Normandy invasion, about a third of their entire munitions output to air defense, according to Burton H. Klein. "It can be seen," he says, "that where the preinvasion attacks really paid off was not nearly so much in the damage they did, but rather in the effect they had on causing the Germans to put a very significant part of their total war effort into air defense." *Germany's Economic Preparations for War* (Cambridge, Harvard University Press, 1959), p. 233.

theoretical formulation of "coercive warfare" might suggest, nevertheless it is worthwhile to sort out some of the different effects and possible purposes of purely destructive exchanges even though there are limits to how neatly a strategy of violence can be tailored to an intent.

One purpose is to intimidate governments or heads of government or to impress them with one's own resolve and one's own refusal to be intimidated. The punitive blow hurts the enemy, implies that more will come unless he desists, and displays resolve or daring in the face of his possible counter-measures. Already this is complex. One can display resolve by hurting oneself, not just by hurting the opponent; and the punitive act can be either an initiative or a response. If a response, it can be conceived and communicated as a "normal" mode of response—just a substitute for tactical military activity elsewhere—or as an "extraordinary" response to some enemy action that is considered to be out of bounds.

The pain and damage could also be aimed at intimidating populations, affecting governments only indirectly. Populations may be frightened into bringing pressure on their governments to yield or desist; they may be disorganized in a way that hampers their government; they may be led to bypass, or to revolt against, their own government to make accommodation with the attacker. Even a few nuclear detonations on a country, unless all news and communication are cut off, would likely dominate civilian life and cause evacuation, absence from work and school, overloading of the telephone system, panic purchas-ing, and various forms of disorder. (If all communications are cut to prevent the news from reaching people and outside radio transmissions jammed for the same purpose, the people may be even more scared.)

Terrorism usually appears to be aimed mainly at intimidating populations and perhaps separating them from their govern-ments. But national leaders can be directly influenced by the prospect of continued pain and destruction, particularly if they are at all responsive to, and part of, the population affected. In many countries, especially in Europe, people have been sensitive to the fact that to be "protected" or "liberated" by a tactical nuclear campaign would hurt. Although local nuclear warfare, whether in Europe or Asia, is usually discussed as though it would be a tactical military campaign, the people in the areas involved are undoubtedly susceptible to nuclear intimidation, and probably so are their leaders. In the event of a tactical nuclear campaign, the outcome might be at least as much affected by the incidental or deliberate civil damage as by the tactical military results. The consequences might be those of nuclear reprisal, on a limited scale, even though the weapons were nominally delivered for tactical purposes.

Limited nuclear exchanges may suddenly look realistic to a decision-maker who is confronted with two familiar and "realistic"alternatives—massive obliterative war and large-scale local defeat. The option of changing the character of war may be what then appears obvious; the idea that one should stick to the local rules and lose tactically when he can shift the war to another basis may be what then looks unreasonable.

There is a more convincing way to make the case that such tactics could occur and may have to be expected if nuclear weapons are used in a local or regional war. It is that this kind of war would grow naturally out of the other kind, the more "tactical" kind of war. Imagine the limited introduction of nuclear weapons for purely tactical purposes in Central Europe. It would be hard in the course of such a war not to notice that a by-product of the tactical use of nuclear weapons was substantial civilian pain and damage and the fear of more. Particularly if one used nuclear weapons to reach a little behind the lines, to destroy rail centers, ports, or air bases, there would be killing of people and collapse of their homes that could not go unnoticed. And there would be fright, refugees clogging the roads in panic and probably reports on both sides that weapons had been used deliberately against towns or cities. Intended or not, such damage and the fear of more would be considerations, possibly decisive, in either side's willingness to continue. Even if not intended, this kind of coercion would be part of what either side was doing in the nominally "tactical" use of nuclear weapons.

Certainly in target selection one would notice that particular targets involved more of this violent by-product than others. If tactically one side were doing well it might bend over backward to keep the war tactically pure and limited, picking targets that minimized nonmilitary damage. But if the other side were doing badly it would certainly recognize that, in the guise of tactical warfare, it could do an enormous amount of punitive damage. If one side hit a few tactical targets that involved disproportionate civilian damage, the other would be under strong temptation to pick a few such targets in return. It is unlikely that when both sides were exchanging sizable punitive attacks as by-products of a tactical campaign they would persist in ignoring a main consequence of their action.

This kind of thing could easily continue under the guise of tactical warfare. Graduated reprisal into the Soviet homeland (or into the West) might take the form of picking nominal targets that were "tactical" or "strategic" in a strictly military sense. But the motivation might become more and more that of subjecting the other side to unbearable punitive pressures, to demonstrate how frightening the war could become, and to intimidate with the threat of further expansion. The fact that the other side can do it, too, would not necessarily dissuade this motivation; and either side, if it wished, could probably persuade itself that the other had been the first to cause unnecessary civilian damage and thus bring the dimension of pure violence into the local war. If the pressure becomes unbearable for both sides, the action may be terminated and a result negotiated that does not reflect the local tactical superiority that one side or the other originally possessed or thought it did.

This has implications for the size and character of military forces, in NATO or elsewhere. If the introduction of nuclear weapons locally and tactically is likely to evolve into an open or disguised war of intimidation, an ability to win the tactical campaign may be neither a necessary condition for success nor a sufficient one.

More important, the purpose of introducing nuclear weapons in a tactical war that one was losing would not be solely, or mainly, to redress a balance on the battlefield. It could be to make the war too painful or too dangerous to continue. Even a limited tactical use of nuclear weapons would be designed to maximize the pressure on the other side to call off the war. And this is less likely to be the pressure of combat casualties and material losses on the fighting front than the pressure of an expanding exchange of limited violence, especially if the war takes place in densely populated parts of the world. These punitive attacks, though they might seem exceedingly slow and measured compared with an all-out attack on the enemy's strategic weaponry, could be fast compared with the pace of tactical warfare. What happens on the battlefield may be of only moderate interest compared with the conduct of such a nuclear war of nerve and endurance. The original theater and   tactics would no longer define either the character or the locale of the war that evolved or the issues involved in its termination.

It would be a mistake to think that conducting war in the measured cadence of limited reprisal somehow rescues the whole business of war from impetuosity and gives it "rational" qualities that it would otherwise lack. True, there is a sense in which anything done coolly, deliberately, on schedule, by plan, upon reflection, in accordance with rules and formulae, and pursuant to a calculus, is "rational" but it is in a very limited sense. It helps if we can slow down a war, induce reflection, and provide national leaders with a consciousness that they are still responsible, still in control, still capable of affecting the course of events. This is different from saying that there is some logical way to conduct a war of limited reprisal or that a decisive intellect can provide sure guidance in such a war on what to do next.

Even if this kind of warfare is irrational it could still enjoy the benefits of slowness, of deliberateness, and of self-control. The situation is fundamentally indeterminate insofar as logic goes. There is no logical reason why two adversaries would not bleed each other to death, drop by drop, each continually feeling that if only he can hold out a little longer the other is bound to give in. There is no assurance that both sides would not come to feel that everything is at stake in this critical test of endurance, that to yield is to acknowledge unconditional submissiveness. It may take luck as well as skill to taper off together in a manner that, leaving neither side a decisive loser in a final showdown of resolve, permits the awful business to come to an end.

Nor is there any guarantee—or even a moderate presumption —that the more rational of two adversaries will come off the better in this kind of limited exchange. There is, in fact, likely to be great advantage in appearing to be on the verge of total abandon. However rational the adversaries, they may compete to appear the more irrational, impetuous, and stubborn.

This is not to deprecate the value of cool, measured, deliberate action in contrast to spasmodic violence. But there are bound to be limits to the safety and security that can be achieved in any style of warfare, if only because limited war is, to a large extent, a competition in the endurance of damage and the acceptance of risk.

China as a "Strategic-Warfare" Adversary

For years most strategic analysts have thought of Communist China itself either as a minor theater in a major war or as an indirect adversary. Hardly anyone seems to have thought about what kind of war it would be or ought to be if the United States became directly engaged with China.

There is little or no visible evidence that in the design of strategic weaponry the possibility of war with China has been taken into account in a manner commensurate with the attention China is now given by the U.S. government. The great diplomatic event of the last half decade is that China is no longer, for purposes of strategic warfare, equivalent to Siberia or the Baltic coast. It is a separate country; and both the Chinese and the Russians have managed to establish that Russia is not absolutely obliged to defend China, or to retaliate on behalf of China, if China should get militarily engaged with the United States. For a long time the idea of a war with China seemed almost meaningless, for although everybody was permitted to express misgivings about American commitments to France and Germany, nobody seemed to have misgivings about the Russian commitment to China. To attack China was merely to give the Russians first strike in a general war. And in that war there was a main adversary, Russia, and all the United States had to do was to obliterate enough of China to destroy the regime, or to satisfy a revenge motive, or to use up whatever weapons it had that could not reach Russia.

Now, however, the overriding consideration in case of engagement with China would be to avoid obliging the Soviet Union to intervene. And if it did not intervene, the ensuing war would have, and should have, almost no similarity to the kind of "general war" that is usually envisaged with respect to the Soviet Union.

The attempt should be to minimize casualties, not to maximize them; there would be no reason to kill Chinese, and there is no historical reason to suppose that the Chinese people, by the hundreds of millions, are any worse threat than any other people except for the regime that heads them in disciplined opposition to us. There may be some reason to threaten to destroy Soviet society in case of general war; I see no reason to threaten to destroy Chinese society in case of general war. There is even less reason actually to destroy Russian society in case of general war, and none at all for destroying Chinese. Somehow the notion got around that the Chinese would still outnumber us if we killed only half of them, and we should therefore try to kill more; this is a grotesque idea and, as far as I can tell, the Chinese do not believe, any more than we should, that victory would go to the side that merely outnumbered the other at the end of a cataclysmic war.

If we did have a war with China, it could be either of two kinds. It could be an effort to destroy the present regime by destroying or disrupting the physical and social basis of its authority and control, with a simultaneous effort to minimize population damage. Or it could be an effort to coerce the regime to come to terms, to pull its troops out of India, to withdraw from Formosa, to disarm itself, or something of the sort. In either case, it is virtually certain that we would not and should not rely on our strategic missiles against China.

We should not because that is probably the most expensive way to destroy the targets that would need to be destroyed and the way least consistent with the constraints we should observe, to wit, minimizing gratuitous population damage, minimizing the Soviet obligation to intervene, and minimizing postwar revulsion against the way we had fought the war. And we would not, because the need to keep our deterrent force intact and ready, to keep the Russians at bay, would be greater than it had ever been before. A war with China would be precisely the time when the United States ought not and would not use a substantial proportion of its strategic deterrent weapons against a second-rate enemy, when Polaris and Minuteman weapons would be valued far beyond their historical money cost.

Furthermore, coercive warfare against Communist China, intended not to destroy the regime but to make the regime behave, would probably be aimed at Chinese military potency and objects of high value to the regime. The two least appropriate, or least effectual, weapons might be the two that people seem readiest to contemplate: conventional explosives and megaton warheads. It might indeed take nuclear weapons to shock the Chinese into an appreciation that we were serious; and it probably would take nuclear weapons, in the face of whatever attrition rate the Chinese could force on us in a protracted campaign, to give us any commanding ability to inflict military or economic damage on them. What the United States was doing in North Vietnam in 1965 against a third-rate adversary, with conventional explosives carried by airplanes that were not designed for the purpose, it would probably attempt to do in China with low-yield nuclear weapons in airplanes that have not yet been designed for it.

We should probably want to destroy the Communist Chinese force with weapons that would cause no casualties beyond a half mile or so from the airfields; we should want to destroy industrial facilities that had a low population or labor-force density. We should want to destroy transport and communication facilities, military depots and training facilities, and troops themselves. We might not afford to do it with conventional weapons (unless newly effective conventional weapons were designed for it) and could not afford to cover such a target system with precious strategic missiles.

We need to recognize that China, as a "strategic" adversary, could not be taken care of by "strategic-war" planning that was developed during two decades of preoccupation with the Soviet Union. China is a different strategic problem altogether. New modes of coercive limited warfare might have to be developed for coping with the problem. The entire tempo of war would be wholly different from anything contemplated against the Soviet Union; except for a small retaliatory force that the Chinese might possess some time in the future, there would be few or no targets of such urgency as to make the initial moments, even the initial days or weeks, as critical as they are bound to be in planning for the contingency of Soviet-American war. The idea of "limited strategic war" between the Soviet Union and the West is often dismissed as plain impracticable, and those who dismiss it may be right; between China and the United States a war would have whatever tempo the United States decided on, or a tempo determined by Chinese actions in some local theater, not the hypersonic tempo of preemptive thermonuclear exchanges.

The need to distinguish a campaign intended to eliminate the regime from one intended only to coerce the regime into good behavior could become supremely important when the Chinese possess a nuclear retaliatory capability (against the United States or against any other population center that they might choose). Making clear to them that, however bad the war already was for them, it could become much, much worse, might be the most effective way to keep that capacity for nuclear mischief disarmed. At the same time, the most potent coercion might be a target strategy that threatened the regime— eventually, gradually, or uncertainly, not suddenly and decisively—and such a strategy would require discriminating what it is that the regime most treasures and where it is most vulnerable.

Whatever its effect on the North Vietnamese willingness to support the Vietcong, and whatever the capacity of North Vietnam to control the Vietcong in submission to the threat of continued bombing attacks, the bombing of North Vietnam must have had one implication for China that went far beyond the war in Southeast Asia. Forcible resistance to them outside their borders can never cost the Chinese more than the resources they knowingly put at risk, the troops and supplies they send abroad; but the bombing of North Vietnam is a mode of warfare that the record now shows to be a real possibility, one that the United States has not only thought of but engaged in. It is a mode of warfare that, at least with air supremacy and the absence of modern anti-aircraft weapons, can be conducted deliberately over a protracted period. And it is a mode of warfare that, if quantitatively increased, could cause extensive physical damage inside the target country, denying any guarantee that the costs of aggression could be confined to the expeditionary force put at risk outside one's border.

Nuclear weapons (or other unconventional weapons) are hardly discussed in connection with the North Vietnamese bombing campaign, presumably because nuclear weapons are not essential to the campaign and because the issues in Southeast Asia are not yet commensurate with the issues raised by nuclear weapons themselves. For translating the North Vietnamese campaign to China, though, nuclear weapons are sure to be considered, not only because their greater efficiency may be more decisive but because the issues involved in a coercive attack on China itself would be correspondingly greater and more likely to equal or to exceed in seriousness the rupturing of our antinuclear traditions. Whether intended or not, the air attack on North Vietnam must carry a warning message to China, a message more credible than the massive threat of megatons on Peking, more potent than the threat of logistical support to the Indians or of Korea-type opposition elsewhere in Asia.

All of this does not mean expecting a war with China, any more than preoccupation with deterrent forces has meant settling for a war with the Soviet Union. It means making sure that *if* the point should be reached where a war with China were contemplated or forced on us, we would not fight a preposterously wrong kind of war for lack of having thought in advance about it or for lack of having equipped ourselves for a major adversary that differs drastically from the adversary that motivated our strategic weapons design for two decades.

It also means thinking about the kind of threat that we wish to pose to the Chinese. There may well be something incredible about a threat to drop megaton weapons on cities like Peking, as well as the threat to engage them in a Korea-type encounter. If they are to be threatened with anything other than rebuff outside their borders, it must be a kind of war that is not wholly incompatible with our principles, with the need to keep the Soviet Union deterred, and with the forces we have available. A major attack on India could make all of this suddenly relevant, just as the Vietnamese war suddenly made relevant a concept of warfare that did not conform to the model of "limited war" that we inherited in Korea.

5
THE DIPLOMACY
OF ULTIMATE SURVIVAL

As a doctrine, "massive retaliation" (or rather, the threat of it) was in decline almost from its enunciation in 1954. But until 1962 its final dethronement had yet to be attempted. All-out, indiscriminate, "society-destroying" war was still ultimate monarch, even though its prerogative to intervene in small or smallish-to-medium conflicts had been progressively curtailed. Beyond some threshold all hell was to be unleashed in a war of attempted extermination, a competition in holocaust, a war without diplomacy and without "options" yet unused, a war in which the backdrop of ultimate deterrence had collapsed on the contenders—a war that would end when all weapons were spent. But in his speech at Ann Arbor, Michigan, in June 1962—a speech reportedly similar to an earlier address in the NATO Council—Secretary McNamara proposed that even in "general war" at the highest level, in a showdown war between the great powers, destruction should not be unconfined. Deterrence should continue, discrimination should be attempted, and "options" should be kept open for terminating the war by something other than sheer exhaustion. "Principal military objectives . . . should be the destruction of the enemy's military forces, not of his civilian population . . . giving the possible opponent the strongest imaginable incentive to refrain from striking our own cities."[1]

The ideas that Secretary McNamara expressed in June 1962 have been nicknamed the "counterforce strategy." They have occasionally been called, as well, the "no-cities strategy." As

1. See above, pp. 24–26.

good a name would be "cities strategy." The newer strategy at last recognized the importance of cities—of people and their means of livelihood—and proposed to pay attention to them in the event of major war.

Cities were not merely targets to be destroyed as quickly as possible to weaken the enemy's war effort, to cause anguish to surviving enemy leaders, or to satisfy a desire for vengeance after all efforts at deterrence had failed. Instead, live cities were to be appreciated as assets, as hostages, as a means of influence over the enemy himself. If enemy cities could be destroyed twelve or forty-eight hours later and if their instant destruction would not make a decisive difference to the enemy's momentary capabilities, destroying *all* of them at once would abandon the principal threat by which the enemy might be brought to terms.

We usually think of deterrence as having failed if a major war ever occurs. And so it has; but it could fail worse if no effort were made to extend deterrence into war itself.

Secretary McNamara incurred resistance on just about all sides. The peace movements accused him of trying to make war acceptable; military extremists accused him of weakening deterrence by making war look soft to the Soviets; the French accused him of finding a doctrine designed for its incompatibility with their own "independent strategic force"; some "realists" considered it impractical; and some analysts argued that the doctrine made sense only to a superior power, yet relied on reciprocity by an inferior power for which it was illogical. The Soviets joined in some of these denunciations and have yet to acknowledge that they share the American government's interest in limiting such a war—though their reaction acknowledges receipt of the message.

This was the first explicit public statement by an important official that deterrence should be extended into war itself and even into the largest war; that any war large or small might have the character of "limited war" and ought to; that (as live captives have often been worth more than enemy dead on the battlefield) live Russians and whole Russian cities together with our unspent weapons might be our most valuable assets, and that this possibility should be taken seriously in war plans and the design of weapons. The idea was not wholly unanticipated in public discussion of strategy; but suggestions by analysts and commentators about limiting even a general war had never reached critical mass. Secretary McNamara's "new strategy" was one of those rare occurrences, an actual policy innovation or doctrinal change unheralded by widespread public debate. Still, it was not altogether new, having been cogently advanced some 2,400 years earlier by King Archidamus of Sparta, a man, according to Thucydides, with a reputation for both intelligence and moderation.

"And perhaps," he said,

> when they see that our actual strength is keeping pace with the language that we use, they will be more inclined to give way, since their land will still be untouched and, in making up their minds, they will be thinking of advantages which they still possess and which have not yet been destroyed. For you must think of their land as though it was a hostage in your possession, and all the more valuable the better it is looked after. You should spare it up to the last possible moment, and avoid driving them to a state of desperation in which you will find them much harder to deal with.[2]

Enemy Forces and Enemy Cities

There were two components of the strategy that Secretary McNamara sketched. Most comment has implied that they are two sides of the same coin, and whether we call it heads or tails we mean the same. But they are distinct. "Counterforce" describes one of them, "cities" (or "no-cities") the other. The two overlap just enough to cause confusion.

Badly expressed they sound alike. In "counterforce" language the principle is to go for the enemy's military forces, not for his cities (not right away, anyhow). In "no-cities" language, the principle is to leave the cities alone, at least at the outset,

2. *The Peloponnesian War,* pp. 58–59.

and confine the engagement to military targets. If we were at a shooting gallery, had paid our fee and picked up the rifle and could shoot either the clay pipes or the sitting ducks, "shoot the pipes" would mean the same as "don't shoot the ducks." But we are not talking about a shooting gallery. The reason for going after the enemy's military forces is to destroy them before they can destroy our own cities (or our own military forces). The reason for not destroying the cities is to keep them at our mercy. The two notions are not so complementary that one implies the other: they are separate notions to be judged on their separate merits.

There is of course the simple-minded notion that war is war and if you are not to hit cities you have to hit something. But that comes out of the shooting gallery, not military strategy. The idea of using enemy cities as hostages, coercing the enemy by the threat of their destruction, can make sense whether or not the enemy presents military targets worth spending our ammunition on.

It may not make sense; the enemy may be crazy, he may not be equipped to know whether or not we have yet destroyed his cities, he may not be able to control his own conduct according to the consequences we confront him with. But if it does make sense, or is worth trying at the outset, it makes sense whether or not we can simultaneously conduct an effective campaign to reduce his military capabilities.

The counterforce idea is not simply that one has to shoot something, and if cities are off limits one seeks "legitimate" targets in order to go ahead with a noisy war. It is a more serious notion: that a good use of weapons is to spend them in the destruction of enemy weapons, to disarm the enemy by trading our weapons for his. If we can forestall his attack on our cities by a disarming attack on his weapons, we may help to save ourselves and our allies from attack.

The "counterforce" idea involves the destruction of enemy weapons so that he cannot shoot us even if he wants to. The "cities" idea is intended to provide him incentive not to shoot us even if he has the weapons to do it. (It can also, with no loss

of manliness, be recognized as a decent effort to keep from killing tens of millions of people whose guilt, if any, is hardly commensurate with their obliteration.)

The two notions complement each other, of course, in that
both are intended to keep the enemy from using his weapons
against us, one through forcible disarmament and the other
through continued deterrence. There is some incompatibility,
though. The city-hostage strategy would work best if the enemy
had a good idea of what was happening and what was not happening, maintained control over his own forces, could perceive
the pattern in our action and its implications for his behavior,
and even were in direct communication with us sooner or later.
The counterforce campaign would be noisy, likely to disrupt the
enemy command structure, and somewhat ambiguous in its
target selection as far as the enemy could see. It might also impose haste on the enemy, particularly if he had a diminishing
capability to threaten our own cities and were desperate to use
it before it was taken away from him.

Nevertheless, a furious counterforce campaign would make
the enemy know there was a war on, that things were not completely under control and that there was no leisure for protracted negotiations. If his cities were to be threatened more
than verbally, so that he knew we meant it, it might be necessary to inflict some damage; doing it in a counterforce campaign that caused a measure of civilian damage might be better
than doing it in a cold-blooded demonstration attack on a few
population centers.

There are, then, different strategies that somewhat support
each other, somewhat obstruct each other, and somewhat compete for resources. Either alone could make sense. A completely
reliable and effective counterforce capability would make it
unnecessary to deter the enemy's use of his weapons by keeping
his cities conditionally alive; it would simply remove his weapons. And a completely successful threat against his cities would
immobilize his weapons and induce capitulation. (In the latter
case the "war" would not look like a big one, in noise and dam-

age, but the sense of commitment and showdown could make it "all-out" in what was at stake.) [3]

The question is often raised whether a counterforce strategy is not self-contradicting: it depends on a decisive military superiority over the enemy and yet to succeed must appeal equally to the enemy, to whom it cannot appeal because he must then have a decisive inferiority. This widespread argument contains a switch between the two meanings, "counterforce" and "cities." A decisive capability to disarm the enemy and still have weapons left over, in a campaign that both sides wage simultaneously, is not something that both sides can exploit. Both may aspire to it; both may think they have it; but it is not possible for both to come out ahead in this contest. (It could be possible for *either* to come out ahead according to who caught the other by surprise. In that case we should say that each had a "first-strike counterforce capability," superiority attaching not to one side or the other but to whoever initiates the war. This is an important possibility but not one the United States government aspired to in its counterforce strategy.)

It can, however, make sense for both sides to take seriously a "cities" strategy that recognizes cities as hostages, that exploits the bargaining power of an undischarged capacity for violence, threatening damage but only inflicting it to the extent necessary to make the threat a lively one. In fact, this "cities" aspect of the so-called "counterforce" strategy should appeal at least as much to the side with inferior strategic forces. If the inferior

3. The nearest the Administration came to making the distinction emphasized here is in an address of John T. McNaughton, General Counsel of the Department of Defense, at an Arms Control Symposium in December 1962. "There is the assertion that *city avoidance* must equal *disarming first strike*" (his italics). "This is wrong. The United States does not think in terms of hitting first. The city-avoidance strategy is no more nor less than an affirmation that, *whatever other targets may be available*" (my italics), "and whoever initiates the use of nuclear weapons, the United States will be in a position to refrain from attacking cities. But it will have in reserve sufficient weapons and it will have the targeting flexibility to destroy enemy cities if the enemy strikes cities first." (This left open the question, raised above in Chapter 4, whether "cities" are an all-or-none category.) *Journal of Conflict Resolution, 7* (1963), 232.

side cannot hope to disarm its enemy, it can survive only by sufferance. It can induce such sufferance only by using its capacity for violence in an influential way. This almost surely means not exhausting a capacity for violence in a spendthrift orgy of massacre, but preserving the threat of worse damage yet to come.

Some commentators calculated that the Soviets would merely "disarm" themselves by directing their weapons at American forces. Having observed that a "counterforce" campaign made no sense they concluded, on the analogy of the shooting gallery, that the Soviets naturally had to fire all their weapons somewhere else. And where else could that be but cities? A facetious answer that brings out the speciousness of the argument is that the Soviets could just as well fire their missiles at their own cities. By firing all their weapons at American cities they virtually guarantee the destruction of their own, and historians would not much care whether the Soviet cities were destroyed by weapons produced domestically or abroad. The idea that restraint in warfare, if it favors the United States, could not be in the Soviet interest has about the same compelling appeal as the idea that a Japanese surrender in 1945, if it favored the United States, could not make sense to the Japanese.[4]

4. Evidently a counterforce campaign that did not destroy cities and populations would require that weapons not be located so near to cities as to merge with them into a single system of targets, and require some protection against radioactive fallout to keep people from being merely destroyed more silently, a little later, by weapons exploded at a distance. The United States conspicuously located Minuteman missiles, for the most part, in the less populous parts of the country, although it did not go to the expense of relocating bomber bases away from the population centers that, for historical reasons, they tended to be close to. It has been argued that the Soviet Union, if it continues to have a numerically inferior missile force and wants to deny the United States a capacity to attack Soviet missiles in a "clean," no-cities war, might choose to keep missiles and bombers near cities; a reciprocated counterforce *and* no-cities war would then be physically impossible, and war might seem less inviting to the United States. A "massive retaliation" would be guaranteed by the lack of any Soviet motive to spare cities and to bargain. If this were done it would not be the first time a government used its own population as a "shield" for its military forces, daring an enemy to do his worst. There are things to be said both for and against the idea—in my opinion

Separating the two components of this strategy is also necessary in dealing with whether a "counterforce" strategy is of transient or enduring interest. There has been a genuine argument whether the United States can reliably expect a capability to disarm the Soviet Union by an offensive campaign, bolstered by defense of the homeland. By "genuine," I mean an argument in which either side could be right depending on the facts and neither can win by sheer logic or casuistry. It is going to depend on technology, intelligence, costs, and the sizes of budgets; and the actual facts may never be reliably clear. By the middle of the 1960s neither side had any clear-cut win in the argument. Testimony of the Defense Department hinted that the United States could not count on a good counterforce capability indefinitely. But if we distinguish the "counterforce" from the "city-threatening" components of the strategy, it is evident that one part of the strategy does, and the other does not, depend on the outcome of this argument. If it is going to turn out as a result of technology, budgets, and weapon choices, that we do not have a capability to disarm the enemy forcibly, then of course a strategy that depends on doing so becomes obsolete— at least until some later time when that capability is available. But there is no reason why that makes the "cities" strategy obsolete. In fact, it virtually yields front rank to the "cities" strategy.

One might pretend, in order to make war as fearsome as possible, that the obvious way to fight a war if we cannot successfully destroy military forces is to destroy the enemy's cities, while he does the same to us with the weapons that we are powerless to stop. But, once the war started, that would be a

___

much more against, even for the Soviet Union—but the point that needs emphasis here is that, though this could frustrate a counterforce city-avoidance campaign, it would not make city destruction a more sensible mode of warfare. It is simply a precarious means of making Soviet weapons less vulnerable by reducing American motives to attack them—confronting the American government with a choice *between* "counterforce" and "city-threatening" strategies, and no opportunity to combine them—and if war should come the motives for restraint should be no less, possibly greater, than if weapons were segregated from people.

witless way to behave, about as astute as head-on collision to preserve the right of way. And general nuclear war is probably fearsome enough anyway to deter any but a most desperate enemy in an intense crisis; making it somewhat less fearsome would hardly invite efforts to test just how bad the war would be. And in the intense crisis, belief that the war could be controlled if it broke out, and stopped short of cataclysm, might actually help to deter a desperate gamble on preemption. So the alleged hard choice between keeping deterrence as harsh as possible and making war, if it should occur, less harsh may not be the dilemma it pretends to be.

The Confrontation of Violence with Violence

The situation in which either side could hurt the other but not disarm it could arise in two different ways. It could arise through both sides' procuring and deploying forces of such a kind that each force is not vulnerable to disarming attack by the other side. Or it could come about through warfare itself.

Discussions of "counterforce warfare" often imply that the war involves two stages. In the first, both sides abstain from an orgy of destruction and concentrate on disarming each other, the advantage going to the side that has the bigger or better arsenal, the better target location and reconnaissance, the advantage of speed and readiness, and the better luck. At some point this campaign is over, for one side or both; a country runs out of weapons or runs out of military targets against which its weapons are any good, or reaches the point where it costs so many weapons to destroy enemy weapons that the exchange is unpromising.

At this stage it is possible, but only barely possible, that both sides have disarmed themselves and each other, and they are momentarily secure from further attack. But any practical evaluation suggests that each side would have weapons left over by the time it had done all the counterforce damage it could do, or could afford to do, to the other's arsenal. Residual weapons will remain and the war is not over.

Now what happens? The more optimistic explanations of a

counterforce strategy imply that at this point the United States has a preponderance of residual weapons, therefore an overwhelming bargaining power, and faces the prospect of an "all-out" city-destruction war with less to lose than the enemy, whose residual arsenal can do some damage while ours can do damage unlimited.This threatened city war is usually implied to be an all-or-none affair, like full-speed collision on the highway, and the driver who has his whole family in the car is expected to yield to the driver who has only part of his family in the car.

This is unsatisfactory. This counterforce exchange—this first stage in major war—accomplishes partial disarmament on both sides, possibly quite unequally, setting the stage in noisy and confusing fashion for a second stage of dirty war, a stage of nuclear bargaining with cities at stake, a stage of "violence," of implicit and explicit threats and likely some competitive destruction of cities themselves to a length that is hard to foretell. Two adversaries face each other in the knowledge that war is on, each capable of large-scale damage, probably unprecedented damage, possibly damage beyond the ability of either to survive with any political continuity. If each retains more than enough to destroy the other, the counterforce exchange was merely a preliminary, a massive military exercise creating great noise and confusion (and undoubtedly great civilian damage too), but constituting an overture to the serious war that is about to begin. If one side is less than able, or each side less than able, substantially to destroy the other, the counterforce stage made a difference but was nevertheless a prelude to the serious exploitation of violence that is about to begin.

So we have two routes that might lead to this confrontation of violence, one by way of procurement and technology in peacetime, the other by way of a counterforce campaign in war itself. The situation would not be this static, of course. It could be that one side can further disarm the other, in a process that takes time, so that there is pressure on the country with the more vulnerable forces to exploit its capacity for violence before it is taken away by the enemy. And of course, if this brink is arrived at through counterforce warfare, the situation is one of fright and alarm, noise and confusion, pain and shock, panic or desperation, not just a leisurely confrontation of two countries measuring their capacities for violence. The two "stages" could overlap—indeed, if counterforce action were unpromising for one side, it might omit that stage altogether and proceed with its campaign of coercion. It might in fact be forced to accelerate its campaign of terror and negotiation by the prospect of losing part of its bargaining power to the other side's counterforce action.

We know little about this kind of violence on a grand scale. On a small scale it occurs between the Greeks and the Turks on Cyprus and it occurred between the settlers and the Indians in the Far West. It occurs in gang warfare, sometimes in racial violence and civil wars. Terror is an outstanding mode of conflict in localized primitive wars; and unilateral violence has been used to subdue satellite countries, occupied countries, or dissident groups inside a dictatorship. But *bilateral* violence, as a mode of warfare between two *major* countries, especially nuclear-armed countries, is beyond any experience from which we can draw easy lessons.

There are two respects in which a war of pure violence would differ from the violence in Algeria or Cyprus. One is that insurgency warfare typically involves two actively opposed sides —the authorities and the insurgents—and a third group, a large population subject to coercion and cajolery. Vietnam in the early 1960s was less like a war between two avowed opponents than like gang warfare with two competing gangs selling "protection" to the population.

There is a second difference. It involves the technology of violence. Most of the violence we are familiar with, whether insurgency in backward areas or the blockade and strategic bombing of World Wars I and II, were tests of endurance over time in the face of violence inflicted over time. There was a limit on how rapidly the violence could be exercised. The dispenser of violence did not have a reservoir of pain and damage that he could unload as he chose, but had some maximum rate of delivery; and the question was who could stand it longest, or who could display that he would ultimately win the contest and so persuade his enemy to yield. Nuclear violence would be more in the nature of a once-for-all capability, to be delivered fast or slowly at the discretion of the contestants. Competitive starvation works slowly; and blockade works through slow strangulation. Nuclear violence would involve deliberate withholding and apportionment over time; each would have a stockpile subject to rapid delivery, the total delivery of which would simply use up the reserve (or the useful targets).

If the Western alliance and the Soviet bloc ever began an endurance contest to see who could force the other to yield through the sheer threat of persistent nuclear destruction, the question would not be who could longest survive some technologically determined rate of destruction but who could most effectively exploit a total capacity the delivery of which was a matter of discretion. If both engaged in a contest of destruction as fast as they were able, each hoping the other would yield first, the destruction might be absolute within a period too short for negotiation. Neither could sensibly initiate maximum destruction, hoping the other side would quit and sue for a truce; time would not permit. Each would have to consider how to measure out its violence. This adds a dimension to the strategy, the dimension of apportionment over time.

What we are talking about is a war of pure coercion, each side restrained by apprehension of the other's response. It is a war of pure pain: neither gains from the pain it inflicts, but inflicts it to show that more pain can come. It would be a war of punishment, of demonstration, of threat, of dare and challenge. Resolution, bravery, and genuine obstinacy would not necessarily win the contest. An enemy's *belief* in one's obstinacy might persuade him to quit. But since recognized obstinacy would be an advantage, displays or pretenses of obstinacy would be suspect. We are talking about a bargaining process, and no mathematical calculation will predict the outcome. If I waylay your children after school, and you kidnap mine, and each of us intends to use his hostages to guarantee the safety of his own children and possibly to settle some other disputes as well, there is no straightforward analysis that tells us what form the bargaining takes, what children in our respective possessions get hurt, who expects the other to yield or who expects the other to expect oneself to yield—and how it all comes out.

There has been remarkably little analysis of this problem in print. It has sometimes been argued, superficially, that the Soviets might blast an American city to prove they meant business, that the Americans would be obliged to blast two Soviet cities in return, and that the Soviets would feel it incumbent on them to blast three (or four?) cities in return, and so the process would go, growing in intensity until nothing was left. This is an important possibility, but there is nothing "natural" about it. It is not necessarily a submissive response to destroy half as much in return rather than twice as much. The appropriate strategy for showing resolve, firmness, endurance, contempt, and righteousness, is not an easy one to determine. The cold-blooded *acceptance* of pain might be just as impressive as the cold-blooded infliction of it. Pericles endorsed the principle when he told the people of Athens, in the face of a Spartan ultimatum, "And if I thought I could persuade you to do it, I would urge you to go out and lay waste your property with your own hands and show the Peloponnesians that it is not for the sake of this that you are likely to give in to them."

This is a strange and repellent war to contemplate. The alternative once-for-all massive retaliation in which the enemy society is wiped out as nearly as possible in a single salvo is less "unthinkable" because it does not demand any thinking. A single act of resignation, however awful the consequences, is still a single act, an exit, not only a resignation *to* one's fate but a resignation *of* responsibility. The suspense is over. And it may seem less cruel because it does not have a cruel purpose—it is merely purposeless—compared with deliberate, measured violence that carries the threat of more. It does not require calculating how to be frightful, how to terrorize an adversary, how to behave in a fearsome way and how to persuade somebody that we are more callous or less civilized than he and can stand the violence and degradation longer than he can. A pure spasm of massive retaliation, if believed to be sufficiently catastrophic, is more like an act of euthanasia, while a consciously conducted "cities strategy" has the ugliness of torture. Maybe one of the reasons why thermonuclear warfare has been likened to "mutual suicide" is that suicide is often an attractive escapist solution compared with having to go on living. Still, though a conscious "cities" strategy may be uglier, it would be more responsible than automated all-out fury.

And it might involve the destruction of some cities. We could hope, of course, that by verbally threatening to destroy all of their cities or none of them, according to whether they surrender or not, we assure their surrender and we both stay alive. If this situation has been arrived at by waging a furious counterforce campaign and possibly a ground war in some theater, there may be a sensation of initiative and desperation that makes the threat more credible than if no war were going on. But what if he does not surrender? If we then have any choice, other than just destroying everything or nothing, we have to think of something impressive in between.

It might be possible to make the massive threat credible by being physically unable to hurt his cities except all at once. If we had a massive bomber fleet approaching his country, capable of devastating his cities in a single attack but incapable of returning home because he could easily destroy the bombers on the ground, equally incapable of staying aloft forever, so that every bomber had to hit a target now or forever lose its ability; and if the time schedule of negotiation made it impossible to halt the attack in mid-course once the first bombs were dropped, the all-or-none threat might be credible—at least if the enemy knows the facts as we know them. Even then he may not comply—or not on the demanding time schedule of bomber sorties—and we must regretfully blow up our influence along with his cities, leaving it his turn to do his worst, or else risk losing all our influence by pulling back altogether. It is about as bad to be thought bluffing as to be bluffing.

So some sheer nuclear violence might be exchanged until the two sides came to terms. There is no guarantee that the terms would reflect the arithmetic of potential violence. If one side can destroy two thirds of the other, and only be destroyed one third itself, this does not guarantee that it wins the bargaining hands down, having its own way altogether just because it is decisively superior. Nor does it mean that it has about "two thirds" of the bargaining power and should expect an outcome with which it is twice as pleased (or half as displeased) as its adversary. There is no simple mathematics of bargaining that tells both sides what to expect so that they can jointly recognize it as a foregone conclusion.

There is no compelling reason to suppose that one side must unconditionally surrender; nor is there any compelling reason to suppose that one side would not unconditionally surrender. The more "successful" the nuclear bargaining is for both sides, the more weapons will be left unused at the end. In the absence of unconditional surrender both sides retain some weapons. (This only means that both sides keep some cities.) It is an inconclusive way to terminate a war, but better than some conclusive ways.

The Crucial Challenge: Ending It

A war of that sort would have to be *brought* to a close, consciously and by design. It could not simply run down by exhaustion when all the targets were destroyed or all the ammunition expended; the whole idea is to keep the most precious targets undestroyed and to preserve weapons as bargaining assets. Some kind of cease-fire or pause would have to be reached and phased into an armistice, by a bargaining process that might at the outset have to be largely tacit, based on demonstration more than on words, but that sooner or later would have to become explicit.

The way the war ended could be more important than the way it began. The last word might be more important than the first strike. There is much preoccupation with the decisive importance of speed and surprise in an initial onslaught, but little attention to what could be equally decisive—a closing stage when the military outcome might be no longer in doubt but the worst of the damage remained to be done. The closing stage, furthermore, might have to begin quickly, possibly before the first volley had reached its targets; and even the most confident victor would need to induce his enemy to avoid a final, futile orgy of hopeless revenge. In earlier times, one could plan the opening moves of war in detail and hope to improvise plans for its closure; for thermonuclear war, any preparations for closure would have to be made before the war starts.

Even the enemy's unconditional surrender might be unavailable unless one had given thought in advance to how to accept and to police surrender. And a militarily defeated enemy, desperate to surrender, might be unable to communicate its offer, to prove itself serious, to accept conditions and to prove compliance on the urgent time schedule of supersonic warfare unless it had given thought before the war started to how it might end. Neither side might be motivated to give the thought, unless there had been at least a tacit understanding that a major war, if it were once started, would also have to be stopped.

It is conceivable that each side would exhaust its weapons in a single-minded counterforce effort to minimize the other's capability to hurt cities. But if each had the ability even to notice what the other was doing, each could see in this process that the other, too, was more concerned to limit damage to itself than to inflict it; so a basis for accommodation would be evident. Furthermore, beyond some point in this military duel, it would make more sense to save weapons and let the enemy shoot at them than to expend them shooting at his. So a sensible war plan ought to provide in advance for the bargaining value of scarce weapons in bringing war to a close and for the likelihood that the enemy, too, would not have spent all his bargaining power in counterforce attacks.

There could be a pause after some initial counterforce onslaught before deliberate, if limited, population attacks began; or there might be no pause. Either way the war must be brought to a close by conscious decisions. It would not stop by itself. Just extending a pause would require decisions to do so.

When and how to stop a war ought to depend upon what the

war is about. But we are talking about what is necessarily a hypothetical war—a war, furthermore, that is universally expected to be so unproductive for either side that even the assumption of a motivating objective may not make sense. The war could have *arisen* over something: Germany, Cuba, Southeast Asia, Middle Eastern oil, the occupation of outer space or the ocean bottom, revolt in a satellite country, assassination, espionage, a false alarm or an accident, even a technological surprise that led one side or the other to launch war in fear for its own security. But even if the war is about something, what it was originally about may soon be swamped by the exigencies of the war itself. Had the Cuban crisis or a Berlin crisis (or even the U-2 crisis, with its Soviet threats to bomb the airfields from which the planes took off) led to a general war, the war would quickly have left Cuba, the Autobahn, or the U-2 launching strips behind.

With today's weapons it is hard to see that there could be an issue about which both sides would genuinely prefer to fight a major war rather than to accommodate. But it is not so hard to imagine a war that results from a crisis' getting out of hand. The aims and objectives, though, on either side would be those of a country that finds itself in a war it did not want and from which it expects no gain.

There may be very few points at which such a war could be stopped. It would be important to identify them ahead of time. Missiles in flight cannot be recalled. (In principle they could be destroyed in flight if it were essential to keep them from reaching their targets, but only if they had been designed to make that possible.) Bombers under radio silence may be beyond recall until their mission is completed. A cloud of radioactive fallout will travel with wind and gravity and cannot be stopped by an armistice. Automated decisions, and plans that make no provisions for stopping in mid-course, may not be susceptible to interruption. Populations would be unwieldy during the process of sheltering. And some episodes in war would be so frantic and confused that messages could not be responded to, or decisions taken, to bring things to a sudden halt.

Different events have different time scales. Missile attacks might be stopped on 30 minutes' notice (corresponding to the time a missile is in flight); bomber attacks might be called off on a few hours' notice if the planes were over enemy territory, more quickly if the planes were not yet in the enemy's air-defense zone. If massive ground attacks are in progress, as they might be if the larger war had grown out of a war in some theater, the time required to halt them would be longer. And if both sides must stop their weapons at approximately the same time, as might be essential and surely would be important, a reciprocally synchronized halt of all important activities would be feasible, at best, only at a few opportune moments, and even then only if both sides were alert to the opportunities and had identified them in advance.

If there were no explicit communication, any pause might depend on one side's stopping its missile launchings and the other side's recognizing that the first had stopped. The signal and its response would consume most of an hour if each side could reliably observe only the impact of enemy missiles, not enemy launching. From the moment one side stops launching, even assuming a clean and sudden stop, it could be upward of 20 minutes before the enemy could notice that missiles had stopped landing; after allowing for reaction time (plus some waiting to make sure) it would be upward of another 20 minutes before the side that first stopped would discern a cessation of impacts in its own country. A lot of war could happen in that much time. A decisive factor would be information; an ability to know that enemy launching has stopped, not just that arrivals on target have stopped, could make a difference to this communication process. And, as is so often the case with deterrence, the enemy's information could be equally important; if we have stopped—if we have initiated a cessation or responded to his —we want him to know it quickly and reliably.

If the war started with a salvo of missiles and the launching of aircraft, there might be an opportunity to stop it before the main bomber fleets of either side had reached their target areas. Missiles would be peculiarly good at hitting quick-moving targets—aircraft on the ground, missiles not yet launched—or for destroying air defenses in order to help the bombers penetrate; the greatest danger to population centers might be forestalled if an armistice could be reached before the bombers arrived. There would be precious little time for negotiations; but already the outcome might be foreseeable, and nothing but inertia or the lack of facilities for reaching a truce would motivate continuation of the war into the bomber stage. (Of course, not all the bombers have to cross oceans to reach their targets; any "stages" would be ragged and approximate.)

Bombers vividly illustrate the dynamic character of war—the difficulty of finding resting places or stopping points, the impossibility of freezing everything. Bombers cannot just "stop." They have to move to stay aloft; to move they have to burn fuel, and while they are moving crews become fatigued, enemy defenses may locate and identify them, and coordination of the bombers with each other will deteriorate. If the planes return to base they have to recycle, with refueling and other delays if the truce was a false one and the war is still on. Finally, at base they may be vulnerable. The aircraft would have been launched quickly for their own safety, and once in the air were immune to missile attack; when they return to base they may be newly vulnerable—if the bases still exist. If they have to seek alternative bases their performance would be further degraded, and so would the threat they pose to the enemy in bringing about, or keeping, a precarious armistice.

Stable stopping points therefore must not only be physically possible, in terms of momentum, gravity, and fuel supplies, and consistent with command arrangements, communications, the speed of decisions, and the information available; they must also be reasonably secure against double cross or resumption of the war.

Armistice and Arms Control

Like any form of arms control, an armistice would probably have to be monitored for compliance. It might possibly take the form of an unnegotiated pause, each side holding fire to see whether the other would too. It would have to be a noticeable pause on both sides, lasting long enough to suggest that it would persist if each conditionally withheld his fire to see if the other did. But, even in this unnegotiated case, both sides would carry out reconnaissance over enemy territory; and once communication were established, negotiation would undoubtedly occur.

There are some good reasons for supposing that, if the war could be stopped, it might be a simple cease-fire that would do it. Considering the difficulty of communication and the urgency of reaching a truce, simple arrangements would have the strongest appeal and might be the only ones that could be negotiated on the demanding time schedule of a war in progress. And a crude cease-fire might be the only stoppage that could be arrived at by tacit negotiation, by the mere extension of a pause. How far the subsequent negotiations could depart from the status quo so established is questionable; neither side would be eager to resume the war, and in default of agreement the thing might just stay stopped. If the cease-fire is only partial, though, or if for reasons like the fuel consumption of loitering aircraft it is inherently unable to endure without more explicit arrangements, the momentary pause would not necessarily determine the main outlines of the final arrangement.

The argument that, of all forms of armistice, the simple cease-fire is the most plausible and likely, is a strong one. Nevertheless, one side or both may not wait for some natural "pause" and may declare or broadcast its willingness to stop and the terms it would expect. (The side first motivated to announce its terms could be either the stronger or the weaker, the one most hurt or the one least hurt, the one with the most yet to lose or the one with the least yet to lose, the one that started the war or the one that did not—and it might not be clear who started it, who had been hurt worse, or who eventually had the most yet to lose.) The natural simplicity of an extended "pause" would depend on actually reaching such a pause; urgency is against it. If verbal exchange can take place, though we should still have to expect any agreement to be crude and simple, it would not have to embody the status quo or even simultaneous stoppage. It does not take long to say "unconditional surrender," and various simple formula might be available if any thought had been given to the matter before the war.

What seems most likely if the war were successfully stopped is a "progressive armistice," with attention going to progressively less urgent things as tentative understandings were reached about the most urgent. First would be the cease-fire itself, which would be sudden if a sudden stoppage is feasible; it would probably not be, so there would have to be some allowance or understanding for weapons and activities already beyond recall. Sanctions or reprisals might have to be threatened for enemy activity that went beyond the tolerance of the agreed cease-fire.

Second would probably be the disposition of residual weapons. An important possibility here would be self-inflicted destruction. If among the terms of the understanding were that one side should disarm itself, partially or completely, of its remaining strategic weapons, ways would have to be found to make it feasible and susceptible of inspection. If it had to be done in a hurry, as it might have to be, enemy aircraft could be required to land at specified airfields, even missiles could be fired at a point where their impact could be monitored (preferably with their warheads disarmed or removed), and submarines might surface to be escorted or disabled. Certainly of all the ways to dispose of remaining enemy weapons, self-inflicted destruction is one of the best; and techniques to monitor it, facilitate it, or even to participate in it with demolition charges would be better than continuing the war and firing away scarce weapons at a range of several thousand miles.

"Uncontested reconnaissance" would be an important part of the process. Submitting to surveillance, restricted or unrestricted, might be an absolute condition of any armistice. In the terminal stage of the war, it is not just "armed reconnaissance" that could be useful but "unarmed reconnaissance," uncontested reconnaissance by aircraft or other vehicles admitted by sufferance.

As in any arms agreement there would be a problem of cheating. And as in any arms agreement, there are two very different dangers. One is that the enemy may cheat and get away with it; the other is that he may not cheat but appear to, so that the arrangement falls apart for lack of adequate inspection. Suppose the armistice is barely one hour old and several nuclear weapons go off in our own country. Has the enemy resumed the war? We may know within another few minutes. Is he testing us, to see how willing we are to resume hostilities, or is he sneaking in a few revenge weapons or perhaps trying to whittle down our postwar military capability? Or was this a submarine or a few bombers that never got the word about the armistice, or confused their instructions and thought they were to carry out their final mission before the armistice? Was this an ally or a satellite of the main enemy, who has not been brought into the armistice? Would the enemy know it if some of his weapons had hit us since the armistice; if we fire a few in reprisal to keep him honest, would he know or believe that these were in response to his own or would he have to assume that we were taking a new initiative, possibly resuming the war? If there is yet no truce covering a European theater, or some overseas bases, how can we tell whether local activity there is a violation of the spirit of the partial truce or merely continued military activity where the truce has not yet been extended?

If questions like these are to be answerable, it will be because they were posed and thought through in advance, recognized at the time the pause or agreement was reached, and even appreciated as pertinent when the weapons themselves were designed and the war plans drawn up. Both sides should want to avoid spoiling the possibility of an armistice through lack of adequate control over their own military forces.

The armistice would induce ambivalent feelings about secrecy. The militarily stronger may be hard-pressed to prove that it is stronger (or to be sure that it is stronger). If one side is submitting to a very asymmetrical disarmament arrangement, it may have to prove how strong it is for bargaining purposes and then prove how weak it is in meeting the disarmament demands of its opponent. For purposes of bluff it would be valuable to have an opponent think one had hidden weapons in reserve; for abiding by a truce arrangement it may be frustrating and dangerous to be unable to deny convincingly the possession of weapons that one actually does not have.

Some Hard Choices

A critical choice in the process of bringing a war to a successful close—or to the least disastrous close—is whether to destroy or to preserve the opposing government and its principal channels of command and communication. If we manage to destroy the opposing government's control over its own armed forces, we may reduce their military effectiveness. At the same time, if we destroy the enemy government's authority over its armed forces, we may preclude anyone's ability to stop the war, to surrender, to negotiate an armistice, or to dismantle the enemy's weapons. This is a genuine dilemma: without technical knowledge of the enemy's command and control system, the enemy's war plan and target doctrine, the vulnerabilities of enemy communications and the procedures for implementing military action, we cannot reach a conclusion here. All we can do is to recognize that there is no obvious answer. Victorious governments have usually wanted to deal with an authority on the other side that could negotiate, enter into commitments, control and withdraw its own forces, guarantee the immunity of ambassadors or surveillance teams, give authoritative accounts of the forces remaining, collaborate in any authentication procedures required to verify the facts, and institute some kind of order in its own country. There is strong historical basis for presuming that we should badly want to be sure that an organized enemy government existed that had the power to demand its armed forces cease, withdraw, submit, mark time, or perform services for us. This has to be weighed against the advantage of disorganizing the initial enemy attacks by destroying the enemy command structure. It may be that there is a clear answer one way or the other, but in this book we do not know which way the answer goes; we know only that it is important. Crudely speaking the questions are whether the enemy's command structure is more vital to the efficient waging of war or the effective restraint and stoppage of war, and which of the two processes is more important to us.

There could of course be a distinction between preserving the political leadership in the enemy country, together with its means of communication and command, and destroying or isolating it while leaving intact the military command structure that might be able to come to terms to stop the war. This is not an easy choice either; one might think the military would be tougher and more amenable to futile sacrifice, while civilian leaders would try to preserve their country. On the other hand, one may suppose that political leaders have less to live for while the military, whatever their attitude toward sacrifice, may be more realistic about the futility of the enterprise and more devoted to what would endure within the country than to the political fortunes of a regime. Here again the answer depends upon expert knowledge; and the answer would not be easy. But some authority within the country would be needed unless the war is to come to an end by sheer exhaustion of weapons alone; and the traditional principle of destroying the enemy's "will to fight" would have to give way before the more important principle of preserving the enemy's "will to survive," his ability to command, and his "will to come to terms." The so-called "will to fight" is a huge metaphor covering the psychology, the bureaucracy, the electronics, the discipline and authority, and the centralization or decentralization of the enemy's military plans. If we are to get any influence out of our enormous capacity for violence, we had better be sure there is some structure capable of being influenced, and capable in turn of bringing the war under control.

A second dilemma arises in the pressure of time that we would want to impose on an enemy. Assuming ourselves militarily ahead after some initial stage, we might find ourselves in the position where a vigorous further prosecution of the war could progressively cut down the enemy's residual forces, and we should have to decide whether that is the most effective way to immobilize his weapons. If we were certain that he would fire all of his weapons as quickly as he could, and fire them to maximize civilian damage on our side, the argument for going after his weapons quickly and unstintingly would be conclusive. If alternatively we were certain that he preferred to pause and negotiate, but nevertheless would fire his weapons rather than see them destroyed on the ground, our all-out attack on them would simply pull the trigger; the argument against it would then be conclusive. These are but two extreme possibilities, but they illustrate how hard the choice would be. An all-out effort *to destroy enemy capabilities* and an all-out effort *to coerce enemy decisions* may not be compatible with each other. There is no conservative way to err on the safe side; we do not know which is the safe side. Next to the choice of preserving or destroying the enemy government, this choice between maximizing enemy weapon-attrition rates or minimizing the urgency to use those weapons may be the most critical, the most difficult, and the most controversial. Here is a point where the distinction between the straightforward application of brute force to block enemy capabilities and the exploitation of potential violence to influence his behavior is a sharp one.

A third choice relates to allied weapons. It is a choice mainly for allied countries that have their own nuclear forces, but it is to some extent a choice the United States could influence. For the next decade or more allied nuclear forces would be of minor significance in the conduct of military operations because they would be few in comparison with our American forces and because our target plans might not be reliably coordinated. Any targets allied countries could destroy, the United States probably feels obliged to attack, too.

If the allied weapons were themselves vulnerable to attack, as aircraft probably would be, they might have to be used quickly to avoid their own destruction; and if their targets were consistent with the American war plan they would merely be used up early in the campaign (and be of little value compared with their cost). If the allied weapons were of such a character that, in addition to being vulnerable to attack, they could be effectively used only against population centers, there would be real danger that they would succeed only in spoiling the prospects for restraint and successful termination. This could virtually make them as much of a threat to American cities (and to their own cities) as to the Soviet cities that they hoped to destroy.

If they were not so vulnerable as to have to fly instantly to target but could be withheld to deter attacks on their own population centers, they might assume rising importance as the war progressed. If the main antagonists, the United States and the Soviet Union, spent a substantial portion of their own weapons in a military duel, the comparative sizes of the allied forces would rise by the mere diminution of the forces they were compared with. The effect this would have on the terminal stage of war would depend critically on how those countries were equipped to participate in any terminal negotiations.

The most successful use of the weapons, from the point of view of the countries concerned, might be to preserve them for continued deterrence, enabling those countries to end the war as nuclear powers. Since these weapons might turn out to have a greater capacity to spoil the American war plan than to contribute to it, this could even be the best employment of the weapons from the American point of view. A strange implication is that, though Europeans have occasionally expressed apprehension that they might end up fighting their own futile war against the Soviet Union while the United States saved itself by keeping its own weapons grounded, an important possibility is exactly the opposite—that they would prefer, if their weapons were not too vulnerable, to count their weapons in the reserve force and let the main expenditure of weapons take place between the two larger military opponents.

Negotiation in Warfare

To think of war as a bargaining process is uncongenial to some of us. Bargaining with violence smacks of extortion, vicious politics, callous diplomacy, and everything indecent, illegal, or uncivilized. It is bad enough to kill and to maim, but to do it for gain and not for some transcendent purpose seems even worse. Bargaining also smacks of appeasement, of politics and diplomacy, of accommodation or collaboration with the enemy, of selling out and compromising, of everything weak and irresolute. But to fight a purely destructive war is neither clean nor heroic; it is just purposeless. No one who hates war can eliminate its ugliness by shutting his eyes to the need for responsible direction; coercion is the business of war. And someone who hates mixing politics with war usually wants to glorify an action by ignoring or disguising its purpose. Both points of view deserve sympathy, and in some wars they could be indulged; neither should determine the conduct of a thermonuclear war.

What is the bargaining about? First there is bargaining about the conduct of the war itself. In more narrowly limited wars—the Korean War, or the war in Vietnam, or a hypothetical war confined to Europe or the Middle East—the bargaining about the way the war is to be fought is conspicuous and continual: what weapons are used, what nationalities are involved, what targets are sanctuaries and what are legitimate, what forms participation can take without being counted as "combat," what codes of reprisal or hot pursuit and what treatment of prisoners are to be recognized. The same should be true in the largest war: the treatment of population centers, the deliberate creation or avoidance of fallout, the inclusion or exclusion of particular countries as combatants and targets, the destruction or preservation of each other's government or command centers, demonstrations of strength and resolve, and the treatment of the communications facilities on which explicit bargaining depends, should be within the cognizance of those who command the operations. Part of this bargaining might be explicit, in verbal messages and replies; much of it would be tacit, in the patterns of behavior and reactions to enemy behavior. The tacit bargaining would involve targets conspicuously hit and conspicuously avoided, the character and timing of specific reprisals, demonstrations of strength and resolve and of the accuracy of target intelligence, and anything else that conveys intent to the enemy or structures his expectations about the kind of war it is going to be.

Second, there would be bargaining about the cease-fire, truce, armistice, surrender, disarmament, or whatever it is that brings the war to a close—about the way to halt the war and the military requirements for stopping it. The terms could involve weapons—their number, readiness, location, preservation, or destruction—and the disposition of weapons and actions beyond recall or out of control or unaccounted for, or whose status was in dispute between the two sides. It would involve surveillance and inspection, either to monitor compliance with the armistice or just to establish the facts, to demonstrate strength or weakness, to assign fault or innocence in case of untoward events, and to keep track of third parties' military forces. It could involve understandings about the reassembling or reconstituting of military forces, refueling, readying of missiles on launching pads, repair and maintenance, and all the other steps that would prepare a country either to meet a renewed attack or to launch one. It could involve argument or bargaining about the degree of destruction to people and property on both sides, the equity or justice of what had been done and the need to inflict punishment or to exact submissiveness. It could involve the dismantling or preservation of warning systems, military communications, or air defenses. And it very likely would involve the status of sheltered or unsheltered population in view of their significance as "hostages" against resumption of warfare.

A third subject of bargaining could be the regime within the enemy country itself. At a minimum there might have to be a decision about *whom* to recognize as authority in the enemy country or with whom one would willingly deal. There might be a choice between negotiating with military or civilian authorities; and if the war is as disruptive as can easily be imagined, there may be a problem of "succession" to resolve. There could even be competing regimes in the enemy country—alternative commanders to recognize as the inheritors of control, or alternative political leaders whose acquisition of control depended on whether they could monopolize communications or get themselves recognized as authoritative negotiators. To some extent, either side can determine the regime on the other side by the process of recognition and negotiation itself. This would espe-

cially be the case in the decision to negotiate about allied countries—China, or France and Germany—or alternatively to refuse to deal with the primary enemy about allied and satellite
affairs and to insist upon dealing separately with the governments of those countries.

A fourth subject for bargaining would be the disposition of
any theater in which local or regional war was taking place.
This could involve the evacuation or occupation of territory,
local surrender of forces, coordinated withdrawals, treatment of
the population, use of troops to police the areas, prisoner exchanges, return or transfer of authority to local governments,
inspection and surveillance, introduction of occupation authorities, or anything else pertinent to the local termination of warfare.

The tempo and urgency of the big war and its armistice might
require ignoring theater affairs in the interest of reaching some
armistice. If so, there might be an understanding, implicit or explicit, that the theater war is to be stopped by unilateral actions
or by immediately subsequent negotiation. There might conceivably be the expectation that the theater war goes on, risking renewed outbreak of the larger war; and possibly the outcome of
the major war would have made the theater war inconsequential
or its local outcome a foregone conclusion. A theater war would
in any case pose acute problems of synchronization: its tempo
would be so slow compared with that of the bigger war that the
terms of the theater armistice simply could not be met within
the time schedule on which the larger war had to be brought to
a close.

Fifth would be the longer term disarmament and inspection
arrangements. These might be comprised in the same package
with the armistice itself; but stopping a war safely and reliably
is different from maintaining safe and reliable military relations
thereafter. The first involves conditions to be met at once, before the war is ended or before planes return to base, before re-
laxation has occurred and before populations have been brought
from their shelters. The second involves conditions to be met
afterward.

For that reason the armistice might, as in the days of Julius Caesar, involve the surrender of hostages as a pledge for future compliance. What form these might take is hard to foretell; but selective occupation of communication centers, preplaced demolition charges, destruction of particular facilities to make a country dependent on outside aid, or even personal hostages might appear reasonable.The purpose of any of these types of hostages—hostages not taken by force but acquired by negotiation—is to maintain bargaining power that would otherwise too quickly disappear. It is to provide a pledge against future compliance, when one's capacity for sanctions is too short-lived. The principle is important, because there is no necessary correspondence between the duration of one's power to coerce and the time span of the compliance that needs to be enforced.

A sixth subject for negotiation might be the political status of various countries or territories—dissolution of alliances or blocs, dismemberment of countries, and all the other things that wars are usually "about," possibly including economic arrangements and particularly reparations and prohibitions. Some of these might automatically be covered in disposing of a theater war; some would already be covered in deciding on the regime to negotiate with. Some might be settled by default: the war itself would have been so disruptive as to leave certain problems no longer in need of solution, certain issues irrelevant, certain countries unimportant.

Of these six topics for bargaining, the first—conduct of the war—is inherent in the war itself if the war is responsibly conducted. The second—terms of armistice or surrender—is inherent in the process of getting it stopped, even though by default most of the terms might be established through an unnegotiated pause. The third—the regime—is at least somewhat implicit in the process of negotiation; the decision to negotiate involves some choice and recognition. The fourth—disposition of local or regional warfare—might be deferred until after the urgent business of armistice had been settled; but the armistice may remain tentative and precarious until the rest of the fighting is actually stopped. The same is probably true of the longer-term disarmament arrangements, and of political and economic arrangements.

We are dealing with a process that is inherently frantic, noisy, and disruptive, in an environment of acute uncertainty, conducted by human beings who have never experienced such a crisis before and on an extraordinarily demanding time schedule. We have to suppose that the negotiation would be truncated, incomplete, improvised, and disorderly, with threats, offers, and demands issued disjointedly and inconsistently, subject to misunderstanding about facts as well as intent, and with uncertainty about who has the authority to negotiate and to command. These six topics are therefore not an agenda for negotiation but a series of headings for sorting out the issues that might receive attention. They are an agenda only for thinking in advance about the termination of war, not for negotiation itself.

How soon should the terminal negotiations begin? Preferably, before the war starts. The crisis that precedes the war would be an opportune time to get certain understandings across. Once war became an imminent possibility, governments might take seriously a "strategic dialogue" that could powerfully influence the war itself. In ordinary peacetime the Soviet leaders have tended to disdain the idea of restraint in warfare. Why not? It permits them to ridicule American strategy, to pose the deterrent threat of massive retaliation, and still perhaps to change their minds if they ever have to take war seriously. On the brink of war they would. It may be just before the outbreak that an intense dialogue would occur, shaping expectations about bringing the war to a close, avoiding a contest in city destruction, and keeping communications open.

It is sometimes wondered whether communications could be established mid-course in a major war. The proper question is whether communications should be cut off. There would have been intense communication before the war, and the problem is to maintain it, not to invent it.

6

THE DYNAMICS
OF MUTUAL ALARM

With every new book on the First World War it is becoming more widely appreciated how the beginning of that war was affected by the technology, the military organization, and the geography of Continental Europe in 1914. Railroads and army reserves were the two great pieces of machinery that meshed to make a ponderous mechanism of mobilization that, once set in motion, was hard to stop. Worse: it was dangerous to stop. The steps by which a country got ready for war were the same as the steps by which it would launch war, and that is the way they looked to an enemy.

No one can quite say just when the war started. There was a great starting of engines, a clutching and gearing and releasing of brakes and gathering momentum until the machines were on collision course. There was no "final" decision; every decision was partly forced by prior events and decisions. The range of choice narrowed until the alternatives were gone.

Railroads made it possible to transport men, food, horses, ammunition, fodder, bandages, maps, telephones, and everything that makes up a fighting army to the border in a few days, there to launch an attack or to meet one, depending on whether or not the enemy got to the border first. Reserve systems made it possible to field an army several times the size that could be afforded continuously in peacetime. Business management on a scale eclipsing any other enterprise known to government or industry determined the railroad schedules, the depots, the order of call-up and shipment, the ratio of horses to caissons, hay to horses, ammunition to gun-barrels, combat troops to field kitchens, the empty cars returning for more, the evacuation of railheads to make room for more troops and kitchens and hay and horses coming in, and the matching of men with units, units with larger units, and the communications to keep them in order.

This miracle of mobilization reflected an obsession with the need for haste—to have an army at the frontier as quickly as possible, to exploit the enemy's unreadiness if the enemy's mobilization was slower and to minimize the enemy's advantages if he got mobilized on the frontier first. The extraordinary complexity of mobilization was matched by a corresponding simplicity: once started, it was not to be stopped. Like rush-hour at Grand Central, it would be fouled up enormously by any suspension or slowdown. A movie of it could be stopped; and while the movie is stopped everything is suspended—coal does not burn in the engines, day does not turn to night, horses get no thirstier, supplies in the rain get no wetter, station platforms get no more crowded. But if the real process is stopped the men get hungry and the horses thirsty, things in the rain get wet, men reporting for duty have no place to go, and the process is as stable as an airplane running out of fuel over a fogged-in landing field. Nor is the confusion merely costly and demoralizing; the momentum is gone. It cannot be instantaneously started up again. Whatever the danger in being slow to mobilize, worse still would be half-mobilization stopped in mid-course.

This momentum of mobilization posed a dilemma for the Russians. The Czar wanted to mobilize against Austria with enough speed to keep the Austrians from first finishing off Serbia and then turning around to meet the threatened Russian attack. The Russians actually had mobilization plans for the contingency, a partial-mobilization plan oriented toward the southern front. They also had full-mobilization plans oriented toward the main enemy, Germany. As a precaution against German attack, full mobilization might have been prudent. But full mobilization would threaten Germany and might provoke German mobilization in return. Partial mobilization against Austria would not threaten Germany; but it would expose Russia to German attack because the partial mobilization could not be converted to full mobilization. The railroads were organized differently for the two mobilization plans. The Russian dilemma was to "trust" in peace with Germany—in the face of a German threat to mobilize if Russia mobilized against Austria—and try to preserve it by mobilizing only against Austria, or to hedge against war with Germany by mobilizing for it and thus to confront Germany with an Eastern enemy mobilizing as though for total war.[1]

How different it would have been if the major countries had been islands, as Britain was. If a hundred miles of rough water had separated every country from its most worrisome enemy the technology of World War I would have given the advantage to the country invaded, not to the invader. To catch the enemy's troop ships on the high seas after adequate warning of the enemy's embarkation, and to fight on the beaches against amphibious attack, with good internal communications and supplies against an enemy dependent on calm seas for getting his supplies ashore—especially for a country that preferred to arm itself defensively, with railroad guns and shore batteries, and submarines to catch the enemy troopships—would have given so great an advantage to the defender that even an aggressor would have had to develop the diplomatic art of goading his opponent into enough fury to launch the war himself. Speed might have mattered to the defender, but not much. If in doubt, wait; or mobilize "partially" until the situation clears up. Being a few days late won't matter if it takes the enemy several days to load his armada and cross the channel; and defensive mobilization will not threaten the other country with attack and provoke its own.

It is not inherent in the logic of warfare, or in the science of weaponry, that haste makes all that difference. With some kinds of geography and technology speed is critical—with other kinds, not. But in 1900, with the transport and military technology

1. See Ludwig Reiners, *The Lamps Went Out in Europe* (New York, Pantheon Books, 1955), pp. 134 ff. His three chapters, 13–15, pp. 123–58, are the best I know on the dynamics of mobilization and their effect on decisions. See also Michael Howard, "Lest We Forget," *Encounter* (January 1964), pp. 61–67.

available then to Europe (and which had been tested in the
Franco-Prussian war), being fast on the draw appeared decisive.

> Victory can only be insured by the creation in peace of an
> organization which will bring every available man, horse,
> and gun (or ship and gun if the war be on the sea) in the
> shortest possible time, and with the utmost possible mo-
> mentum, upon the decisive field of action. . . . The statesman
> who, knowing his instrument to be ready, and seeing war
> inevitable, hesitates to strike first, is guilty of crime against
> his country.

So reads Colonel Maude's introduction to Clausewitz.[2]

Even if we have no control over the way technology unfolds
we can still know what we like. And what we like is a military
technology that does not give too much advantage to haste. We
like that whether we are Russians, Americans, or anybody else.
The worst military confrontation is one in which each side
thinks it can win if it gets the jump on the other and will lose if
it is slow. Let us modify Colonel Maude's statement: The
statesman who, knowing his instrument to be ready *on condition he strike quickly,* knowing the enemy instrument to be
equally ready, knowing that if he hesitates he may lose his instrument and his country, knowing his enemy to face the same
dilemma, and seeing war not inevitable but a serious possibility,
who hesitates to strike first is—what?

He is in an awful position. It is a position that both he and
his enemy can equally deplore. If neither prefers war, either or
both may yet consider it imprudent to wait. He is a victim of a
special technology that gives neither side assurance against attack, neither such a clear superiority that war is unnecessary,
and both sides a motive to attack, a motive aggravated by the
sheer recognition that each other is similarly motivated, each
suspicious that the other may jump the gun in "self-defense."

Among all the military positions that a country can be in, in

2. Karl von Clausewitz, *On War* (New York, Barnes & Noble, 1956), introduction
by F. M. Maude. The date of this introduction is apparently around 1900.

relation to its enemy, this is one of the worst. Both sides are trapped by an unstable technology, a technology that can convert a likelihood of war into certainty. Military technology that puts a premium on haste in a crisis puts a premium on war itself. A vulnerable military force is one that cannot wait, especially if it faces an enemy force that is vulnerable if the enemy waits.

If the weapons can act instantaneously by the flip of a switch, a "go" signal, and can arrive virtually without warning to do decisive damage, the outcome of the crisis depends simply on who first finds the suspense unbearable. If the leaders on either side think the leaders on the other are about to find it unbearable, their motive to throw the switch is intensified.

But almost certainly there is more to it than just throwing the switch; there are things to do, and there are things to look for. Things to look for are signs of whether the enemy is getting closer to the brink or has already launched his force. The things to do are to increase "readiness." Readiness for what?

Some steps can increase readiness to launch war. Some steps reduce vulnerability to attack. The mobilization systems of continental countries in 1914 did not discriminate. What one did to get ready to meet an attack was the same as what one did to launch an attack. And of course it looked that way to the enemy.

There is bound to be overlap between the steps that a country can take to get ready to start a war and the steps it can take to make war less inviting to its enemy or less devastating to itself. There is no easy way to divide the measures of alert and mobilization into "offensive" and "defensive" categories. Some of the most "defensive" steps are as important in launching a war as in awaiting enemy attack. Sheltering the population, if shelter is available, is an obviously "defensive" step if the enemy may launch war before the day is out. It is an equally obvious "offensive" step if one expects to launch an attack before the day is out and wants to be prepared against counterattack and retaliation. To stop training flights and other incidental air force activity, readying the maximum number of bombers on airfields, is a way of assuring greater reprisal against the enemy in case he attacks us; it can also be a step toward readiness to attack the enemy.

Still, though there is overlap, there can be a difference. One readiness step that was widely reported at the time of the Cuban crisis was the dispersal of bombers to alternate airfields. The airfields of many large cities are capable of handling air force bombers; in peacetime it would be a nuisance, an expense and possibly a danger, to keep bombers with bombs dispersed to large-city airfields. But in a crisis, when it is important not to confront the enemy with a bomber force that is too easy a target for his missiles, doubling or trebling the number of bases among which the bombers are dispersed can be worth some nuisance, some expense, even some danger. The bombers are in no better condition to launch an attack if they are dispersed away from their main bases; they may actually be somewhat less ready for a coordinated surprise attack, especially since they may be more susceptible to enemy surveillance. But they are less vulnerable to enemy attack. Thus the *comparison* of our readiness for a war that *we* start and our readiness for a war that the *enemy* starts is changed by such dispersal. Whatever the wisdom of converting large-city airfields into urgent military targets—and it is preposterous unless the bombers are desperately in need of a modest improvement in their security—one can at least recognize that such dispersal mainly reduces vulnerability to attack rather than increasing the advantage to be gained by launching an attack.

There can also be a difference in the sheer timing of mobilization. The enemy can presumably take steps for his own readiness at the same time we take steps for our own. If the steps he takes reduce his vulnerability to attack, reducing the advantage to us of a sudden surprise launch of our strategic forces and giving him greater assurance of our unlikelihood to do that, then just allowing him time for such increased readiness will reduce our offensive capability relative to our defensive, or our "counterforce" capability relative to our "retaliatory" capability. The way both sides alert their forces and mobilize in a crisis can have much to do with whether the situation becomes increasingly dangerous or not. The degree of readiness, the extent of mobilization, the high alert status of strategic forces and a sense of "confrontation" will make the situation tense and expectant and hostile in appearance. The situation may not be more dangerous at the end of a day's mobilization, though, if each side provides the enemy less to be gained by sudden attack and the penalty on waiting (the premium on haste) is reduced.

The Mischievous Influence of Haste

The premium on haste—the advantage, in case of war, in being the one to launch it or in being a quick second in retaliation if the other side gets off the first blow—is undoubtedly the greatest piece of mischief that can be introduced into military forces, and the greatest source of danger that peace will explode into all out war. The whole idea of accidental or inadvertent war, of a war that is not entirely intended or premeditated, rests on a crucial premise—that there is such an advantage, in the event of war, in being the one to start it and that each side will be not only conscious of this but conscious of the other's preoccupation with it. In an emergency the urge to preempt—to preempt the other's preemption, and so on ad infinitum—could become a dominant motive if the character of military forces endowed haste and initiative with a decisive advantage. It is hard to imagine how anybody would be precipitated into full-scale war by accident, false alarm, mischief, or momentary panic, if it were not for such urgency to get in quick. If there is no decisive advantage in striking an hour sooner than the enemy and no disadvantage in striking an hour later, one can wait for better evidence of whether the war is on. But when speed is critical the victim of an accident or a false alarm is under terrible pressure to get on with the war if in fact it is war or if the enemy seems likely, even in "self-defense," to anticipate war by starting it. If each side imputes similar urgency to the other, the urgency is aggravated.

It is not accidents themselves—mechanical, electronic, or human—that could cause a war, but their effect on decisions. Accidents can trigger decisions, and this may be all that anybody has ever meant; but the distinction needs to be made. The remedy is not just preventing accidents, false alarms, or unauthorized ventures, but tranquilizing the decisions. The accident-prone character of strategic forces—more correctly, the sensitivity of strategic decisions to possible accidents or false alarms—is closely related to the security of the forces themselves. If a country's retaliatory weapons are reasonably secure against surprise attack, preemptive or premeditated, the country need not respond so quickly to alarms and excursions. Not only can one wait and see but one can assume that the enemy himself, knowing that one can wait and see, is less afraid of a precipitate decision, less tempted toward a precipitate decision of his own.

But there are two ways to confront the enemy with retaliatory forces that cannot be destroyed in a surprise attack. One is to prevent surprise; the other is to prevent their destruction even in the event of surprise.

Radar, satellite-borne sensory devices to detect missile launchings, and alarm systems that signal when a country has been struck by nuclear weapons, could give us the minutes we might need to launch most of our missiles and planes before they were destroyed on the ground. If the enemy knows that we can react in a few minutes and that we will have the few minutes we need, he may be deterred by the prospect of retaliation. But hardened underground missile sites, mobile missiles, submarine-based missiles, continually air-borne bombs and missiles, hidden missiles and aircraft, or even weapons in orbit do not so much depend on warning; they are designed to survive an attack, not to anticipate it by launching themselves at the enemy in the few minutes after warning—perhaps ambiguous warning—is received. In terms of ability to retaliate, warning time and survivability are to some extent substitutes but they also compete with each other. Money spent dispersing and hardening missile sites or developing and building mobile systems could have been spent on better warning, and vice versa.

More important, they conflict in the strategy of response. The critical question is, what do we do when we do get warning? The system that can react within fifteen minutes may be a potent deterrent, but it poses an awful choice whenever we think we have warning but are not quite sure. We can exploit our speed of response and risk having started war by false alarm, or we can wait, avoiding an awful war by mistake but risking a dead retaliatory system if the alarm was real (and possibly reducing our deterrence in a crisis if the enemy knows we are inclined to give little credence to the warning system and wait until his bombs have landed).

The problem may be personal and psychological as well as electronic; the finest products of modern physics are of no avail if the top ranking decision-maker, whoever he may be within the time available, is too indecisive, or too wise, to act with the alacrity of an electronic computer.

We get double security out of the system that can survive without warning: the enemy knowledge that we can wait in the face of ambiguous evidence, that we can take a few minutes to check on the origin of accidents or mischief, that we are not dependent on instant reaction to a fallible warning system, may permit the enemy, too, to wait a few minutes in the face of an accident and permit them in a crisis to attribute less nervous behavior to us and to be less jumpy themselves. (If we think the other side is taking Colonel Maude's advice, we have an extra reason for taking it ourselves!)

If we think of the decisions as well as the actions we can see that accidental war, like premeditated war, is subject to deterrence. Deterrence, it is often said, is aimed at the rational calculator in full control of his faculties and his forces; accidents, it is said, may trigger war in spite of deterrence. "The operation of the deterrence principle in preventing war," says Max Lerner, "depends upon an almost flawless rationality on both sides." [3] But it is really better to consider the more "accidental" kind of

3. *The Age of Overkill*, p. 27. Incidentally, when people say that "irrationality" spoils deterrence they mean—or ought to mean—only particular brands of it. Leaders can be irrationally impetuous or irrationally lethargic, intolerable of suspense or incapable of decision. A Hitler may be hard to deter because he is "irrational," but a Chamberlain is equally irrational and especially easy to deter. The human inability to rise to the occasion may sometimes lead to a Pearl Harbor, or to a remilitarization of the Rhineland; it probably also cushions a good many shocks, accidents, and false alarms and helps governments to rationalize their way out of crises. This is no consolation when we confront the wrong kinds of madness; still, we may as well get the theory straight.

war—the war that arises out of inadvertence or panic or misunderstanding or false alarm, not by cool premeditation—as *the* deterrence problem, not a separate problem and not one unrelated to deterrence.

We want to deter an enemy decision to attack us—not only a cool-headed, premeditated decision that might be taken in the normal course of the Cold War, at a time when the enemy does not consider an attack by us to be imminent, but also a nervous, hot-headed, frightened, desperate decision that might be precipitated at the peak of a crisis, that might result from a false alarm or be engineered by somebody's mischief—a decision taken at a moment when sudden attack by the United States is believed a live possibility.

The difference is in the speed of decision, the information and misinformation available, and the enemy's expectations about what happens if he waits. The enemy must have some notion of how much he would suffer and lose in a war he starts, and of how much more he may suffer and lose in a war that, by hesitating, he fails to start in time. He must have some notion of how probable it is that war will come sooner or later in spite of our best efforts and his to avert it. In case of alarm he has some estimate, or guess, of the likelihood that war has started and of the risks of waiting to be sure. In deciding whether to initiate war or to respond to what looks like war the enemy is aware not only of retaliation, but of the likelihood and consequences of a war that he does not start, one that we start. Deterring premeditated war and deterring "accidental war" differ in those expectations—in what the enemy thinks, at the moment he makes his decision, of the likelihood that alarms are false ones or true, and of the likelihood that if he abstains, we won't.

Accidental war therefore puts an added burden on deterrence. It is not enough to make a war that he starts look unattractive compared with no war at all; a war that he starts must look unattractive even as insurance against the much worse war that—in a crisis, or after an accident, or due to some mischief, or in misapprehension of our intent—he thinks may be started against him or has already started. Deterrence has to make it never appear *conservative* to elect, as the lesser danger, preemptive war.

"Accidental war" is often adduced as a powerful motive for disarmament. The multiplication and dispersion of ever more powerful weapons seems to carry an ever growing danger of accidental war; and many who are confident that deliberate attack is adequately deterred are apprehensive about the accidental-war possibilities inherent in the arms race.

But there is a conflict, and a serious one, between the urge to have fewer weapons in the interest of fewer accidents and the need—still thinking about "accidental war"—to have forces secure enough and so adequate in number that they need not react with haste for fear of not being able to react at all, secure enough and so adequate in number that, when excited by alarm, we can be conservative and doubt the enemy's intent to attack, and that the enemy has confidence in our ability to be calm, helping him keep calm himself. A retaliatory system that is inadequate or insecure not only makes the possessor jumpy but is grounds for the enemy's being jumpy too.

It is important to keep in mind, too, that (as in any other business) accidents and mischief and false alarms can be reduced by spending more money. To correlate weapons, accidents, and arms budgets ignores the fact that the security of retaliatory forces, the control over them and communication with them, is an important and expensive part of the military establishment. For a given number of weapons, more money may mean more reliable communications and command procedures. Skimpy budgets may mean skimpy protection against malfunction, confusion, and mischief.

Even numbers can help. Few people have kind words in print for "overkill," but it is probably a valid principle that restraining devices for weapons, men, and decision processes—delaying mechanisms, safety devices, double-check and consultation procedures, conservative rules for responding to alarms and communication failure, and in general both institutions and mechanisms for avoiding an unauthorized firing or a hasty reaction to untoward events—can better be afforded, and will be afforded, if there is redundancy in numbers. If weapons are scarce, every restraining device will meet with the argument that some weapons somewhere will fail to get the word, that some lock will be unopened when a weapon should be fired, and that delay will cause some weapons to be fired too late. The best answer to this argument is that there is enough ammunition to keep a few duds from making all that difference and we can afford an occasional malfunction resulting from conservative procedures and restraining devices.

To say this does not prove that a larger strategic force will be less susceptible to accidental or unauthorized launch. But it can be; and while the argument is not of enough weight to pretend to settle the question of disarmament, it surely is of enough weight to be taken into account.

"Vulnerability" and Deterrence

"Vulnerability" is the problem that was dramatized by Sputnik in 1957 and by Soviet announcements then that they had successfully tested an ICBM. Nobody doubted that the aircraft of the Strategic Air Command, if launched against Soviet Russia, could do enormous damage to that country, unquestionably enough to punish any aggression they had in mind and enough to deter that aggression if they had to look forward to such punishment. But if the Soviets were about to achieve a capability to destroy without warning the massive American bomber force while the aircraft were vulnerably concentrated on a small number of airfields, the deterrent threat to retaliate with a destroyed bomber force might be ineffectual. The preoccupation with vulnerability that began in 1957 or so was not with the vulnerability of women and children and their means of livelihood to sudden Soviet attack on American population centers. It was the vulnerability of the strategic bomber force.

This concern with vulnerability led to the improved alert status of bombers so that radar warning of ballistic missiles would permit the bombers to save themselves by taking off. And it led to the abandonment of "soft," large, liquid-fueled missiles like the Atlas, and the urgent substitution of Minuteman and Polaris missiles which, in dispersed and hardened silos or in hidden submarines, could effectively threaten retaliation. An Atlas missile could retaliate as effectively as several Minutemen, if alive, but could not so persuasively threaten to stay alive under attack. In the late 1950s and the early 1960s the chief criterion for selecting strategic weapon systems was invulnerability to attack, and properly so. Vulnerable strategic weapons not only invite attack but in a crisis could coerce the American government into attacking when it might prefer to wait.

Vulnerability was a central theme of the Geneva negotiations in 1958 about measures to safeguard against surprise attack. There is nothing especially heinous about a war begun in surprise; if people were going to be killed it would be small consolation to have the bad news a little before it happened. What made surprise attack a worthy category for consideration in a disarmament conference was precisely this character of strategic weapon systems, the possibility that "surprise" might help an attack to succeed, and by inviting success spoil deterrence. But success would be measured by how well the surprise attack could forestall retaliation on the country launching attack; the measure of success would not be the speed with which cities could be destroyed but the likelihood that the victim's strategic weapons could be destroyed. If enemy bombers could be caught on the ground, with speed and surprise, the enemy population could be disposed of at leisure. Measures that might spoil surprise, or that might make strategic weapons less vulnerable to surprise, if available to both sides and possibly arising out of collaboration between them, might stabilize deterrence and make it more reliable, assuring each side against being attacked and thus reducing each side's incentive to attack.

So we have the anomaly of a great disarmament conference devoting itself in large measure to the protection not of women and children, noncombatants and population centers, but of weapons themselves. If an "open skies" arrangement could make bombers and missiles more secure, keeping the threat of retaliation a lively one no matter who launched the war, the women and children would be safer, not because they would have warning if the war were to come but because the war would be less likely to come. If a city has a limited number of bullet-proof vests it should probably give them to the police, letting the people draw their security from a police force that cannot be readily destroyed.

The Charact of Weapons: Strength vs. Stability

There is, then, something that we might call the "inherent propensity toward peace or war" embodied in the weaponry, the geography, and the military organization of the time. Arms and military organizations can hardly be considered the exclusively determining factors in international conflict, but neither can they be considered neutral. The weaponry does affect the outlook for war or peace. For good or ill the weaponry can determine the calculations, the expectations, the decisions, the character of crisis, the evaluation of danger and the very processes by which war gets under way. The character of weapons at any given time determines, or helps to determine, whether the prudent thing in a crisis is to launch war or to wait; it determines or helps to determine whether a country's preparations to receive an attack look like preparations for attack itself; it determines or helps to determine how much time is available for negotiation on the brink of war; and it determines or helps to determine whether war itself, once started, gets altogether out of control or can be kept responsive to policy and diplomacy.

To impute this influence to "weaponry" is to focus too narrowly on technology. It is weapons, organization, plans, geography, communications, warning systems, intelligence, and even beliefs and doctrines about the conduct of war that together have this influence. The point is that this complex of military factors is not neutral in the process by which war may come about.

Obviously this is so in a one-sided sense. The weak are unlikely to attack the strong, and nearly everybody acknowledges that there is something to "deterrence." This is not what I have in mind; the matter would be simple if relative strength were all that mattered and if relative strength were easy to evaluate. Either the strong would conquer the weak or the strong, if peaceful, would be safe against weaker enemies; combinations might form to achieve a balance or a preponderance, but we would be dealing with simple quantities that could be added up. When I say, though, that "weaponry" broadly defined is an influential factor itself, I refer to its character, not its simple quantity. A military complex cannot be adequately described by a quantity denoting "strength."

One critical characteristic has just been discussed—the dependence on speed, initiative, and surprise. This is different from "strength." If one airplane can destroy 45 on an airfield, catching the other side's airplanes on the ground can be decisively important while having *more* airplanes than the other side is only a modest advantage. If superiority attaches to the side that starts the war, a parade-ground inventory of force—a comparison of numbers on both sides—is of only modest value in determining the outcome. Furthermore, and this is the point to stress, the *likelihood* of war is determined by how great a reward attaches to jumping the gun, how strong the incentive to hedge against war itself by starting it, how great the penalty on giving peace the benefit of the doubt in a crisis.

The dimension of "strength" is an important one, but so is the dimension of "stability"—the assurance against being caught by surprise, the safety in waiting, the absence of a premium on jumping the gun.[4]

4. If not already acquainted with it, the reader should certainly see Albert Wohlstetter's classic, "The Delicate Balance of Terror," *Foreign Affairs, 37* (1959), 211–34; it marks the watershed in professional treatment of the "vulnerability" problem and the stability of deterrence. Malcolm Hoag, "On Stability in Deterrent Races," *World Politics, 13* (1961), 505–27, is a lucid theoretical treatment that contrasts alternative arms technologies and the types of arms race they can generate. T. C. Schelling and Morton H. Halperin consider the arms-control implications in *Strategy and Arms Control* (New York, Twentieth Century Fund, 1961), especially Chapters 1, 2, and 5.

Stability itself has both a static and a dynamic dimension. The static dimension reflects the expected outcome, at any given moment, if either side launches war. The dynamic dimension reflects what happens to that calculation if either side or both sides should *move* in the direction of war, by alert, mobilization, demonstration, and other actions that unfold over time. It involves the steps taken in a crisis. Do we become more vulnerable or less vulnerable as we ready ourselves for the possibility of war, and does the enemy become less vulnerable or more vulnerable and less or more obsessed with his own vulnerability and his need to attack quickly? Equally important: what happens tomorrow and the day after as a result of the steps we take today? If we make ourselves less vulnerable today is it at the expense of tomorrow?

A vivid example of this dynamic problem is bomber aircraft. In case of warning they can leave the ground. If they leave the ground they should initially proceed as though to target; in case it is war, they should not be wasting time and fuel by loitering to find out what happens next. As they proceed to target, they can be either recalled or confirmed on their mission. (The actual procedure may be that they return to base unless confirmed on their mission, by "positive control" command procedures.) If recalled, however, they return to the relative vulnerability of their bases. They need fuel, their crews are tired, they may need maintenance work, and they are comparatively unsynchronized. They are, in sum, more vulnerable, and less ready for attack, than before they took off.

This is a dynamic problem, involving the pressure of time; it is a situation that cannot be sustained indefinitely. It is not an unsolvable problem; but it is one that has to be solved. Like the railroad mobilization of World War I, the bomber arrangements may enjoy simplicity and efficiency by ignoring the possibility that they may have to loiter or return to base. Like the railroad mobilization of World War I, the procedures may coerce decisions unless the procedures are compromised to facilitate orderly return to base. Decisions may be compromised in either of two directions. The planes may fail to take off when they ought to, because of the high cost of spoiling the force on a false alarm and having to return to base disorganized. Or a decision to proceed with war may be coerced by a situation in which aircraft are momentarily in a good position to continue with war and in a poor one to call it off.[5]

If both sides are so organized, or even one side, the danger that war in fact will result from some kind of false alarm is enhanced. This is one of those characteristics of armed forces that influences the propensity toward war and that is not comprised within a calculation of "strength." The Strategic Air Command has undoubtedly been cognizant of this problem and has taken steps to minimize it; the point here is simply that the steps are necessary, they undoubtedly cost something, and the technology of aircraft affects how well the problem can be solved. If the problem is not perceived at the time when the aircraft are designed, or at the time the runways and refueling facilities are provided, the solution of the problem may be less complete or more costly.

The fueling of missiles could have created a similar problem if solid-fueled missiles had not so quickly replaced the originally projected missiles utilizing refrigerated fuels. If it takes *time* to fuel a missile, fifteen minutes or an hour, and if a fueled missile

5. Roberta Wohlstetter, whose unique study of *Pearl Harbor: Warning and Decision* (Stanford, Stanford University Press, 1962), dissected the problem of intelligence evaluation in a crisis, has recently pointed out the crucial interaction between intelligence and response. "In the Cuban missile crisis," she says, "action could be taken on ambiguous warning because the action was sliced very thin. . . . If we had had to choose only among much more drastic actions, our hesitation would have been greater. The problem of warning, then, is inseparable from the problem of decision. . . . We can improve the chance of acting on signals in time to avert or moderate a disaster . . . by refining, subdividing and making more selective the range of responses we prepare, so that our response may fit the ambiguities of our information and minimize the risks both of error and of inaction." "Cuba and Pearl Harbor," *Foreign Affairs, 43* (1965), 707. For an example of action sliced so appallingly thick that paralysis was guaranteed, see Henry Owen's discussion of the Rhineland crisis of 1936, "NATO Strategy: What Is Past Is Prologue," in the same issue, pp. 682–90.

cannot be held indefinitely in readiness, a problem very much like the bomber problem can arise. To fuel a missile is not a simple act of prudence, achieving enhanced readiness at the cost of some fuel that may be wasted and some potential maintenance work on the missiles themselves after the crisis is over. If the fuel begins to dissipate, or the fueled missile becomes susceptible to mechanical fatigue or breakdown, getting a missile ready requires a risky decision. The risk is that the missile will be less ready, after a brief period, than if it had never been made ready in the first place. It, too, like the aircraft burning fuel in the air, can coerce a decision; it can coerce a decision in favor of war once it is fueled and ready and threatens to become unready shortly. It can coerce a decision to remain unready by making it dangerous to put the missile into its mobilization process.

In the mid-1960s, American strategic weapon systems did not appear to have much in common with the mobilization process of 1914. Secure yet quick-firing missiles of the Minutemen and Polaris type, and carefully designed alert procedures for the bombers, appeared to minimize the constraint or coercion on decisions in a crisis. The strategic weapon systems seemed to have a minimum of "dynamic instability" embodied in their alert and mobilization procedures.

Some observers thought this was a disadvantage, because the enemy could not be so readily coerced by American demonstrations, by getting ourselves in a position of temporarily increased readiness, by taking steps that showed our willingness to risk war and that actually increased the risk of war. There were some who thought that bombers were more usable in a crisis than instantly ready missiles, because they could dramatically take off, or disperse themselves to civilian bases, giving an appearance of readiness for war.

They could be right. What needs to be recognized is that the flexing of muscles is probably unimpressive unless it is costly or risky. If aircraft can take off in a crisis with great noise and show of activity, but at no genuine risk to themselves and at modest cost in fuel and personnel fatigue, it may demonstrate

little. The impressive demonstrations are probably the dangerous ones. We cannot have it both ways.[6]

Mobilization: A Contemporary Example

There is nevertheless an important area of mobilization, one little recognized and much underrated, that could prove enormously important in a crisis, for good or ill—for good if one
wants demonstrations, for ill if one does not want to put national decision-makers under acute pressure for a decision, especially for ill if it has not been foreseen and taken into account.
This is the area of civil defense.

Civil defenses are often called "passive defenses," while
anti-missile missiles, anti-aircraft missiles, and interceptor aircraft are called "active defenses." In an important sense,
though, giving the words their ordinary meanings, it is the civil
defenses that are probably the most active and the "active defenses" that would be the most passive. If we should install anti-
missile missiles around our population centers they would
probably be quick-reacting missiles themselves, in a state of
fairly continuous readiness, involving no dramatic readiness
procedures and not being utilized unless threatening objects appeared overhead. One can imagine other kinds of defenses
against ballistic missiles that did involve readiness procedures,
that required decisions to mobilize in advance; perhaps short-
lived orbiting systems that had to be launched in an emergency
in anticipation of attack would have this character. But the systems currently under discussion or development appear to be
relatively "passive." They would sit still in constant readiness

6. Alfred Vagts has a rich chapter on "Armed Demonstrations," in his *Defense
and Diplomacy* (New York, King's Crown Press, 1956). He warns, cogently citing
Disraeli and Churchill on his side, against the demonstration that falls short of
the mark and signals the opposite of stern intent. He believes, too, that a fundamental
change has taken place in "this instrument of diplomacy" in the last thirty years,
namely, "Much if not most of Western demonstrativeness is inward, rather than
outward. It is directed toward their own citizenry, rather than at the address of
the Russians." Whether or not he would change his emphasis today, ten years
later, the point is a valid one.

and fire only in response to the local appearance of hostile objects overhead.

The civil defenses would be a dramatic contrast. Shelters work best if people are in them. The best time to get people in the shelters is before the war starts. To wait until the enemy has launched his ballistic missiles (if one expects some of them to be aimed at cities) would be to leave the population dependent on quick-sheltering procedures that had never been tested under realistic conditions. Even if the enemy were expected not initially to bring any of our cities under attack, fallout from target areas could arrive in periods ranging from, say, a fraction of an hour up to several hours, and in the panic and confusion of warfare a few hours might not be enough. Furthermore, the most orderly way to get people into shelters, with families assembled, gas and electricity shut off, supplies replenished and fire hazards reduced, the aged and the sick not left behind, and panic minimized, would be by sheltering before the war started.

And that means sheltering before war is a certainty. There is a dilemma right here. If sheltering will be taken as a signal that one expects war and intends to start it, sheltering gives notice to the other side. Surprise would depend on not sheltering. A nation's leaders must decide whether the advantage of surprise against the enemy is worth the cost of surprising their own population unprepared. This would be a hard choice. Can one afford to warn his own population if it means warning the enemy? Can one afford surprising the enemy if it means surprising one's own country?

It is unlikely that sheltering would be an all-or-none operation. Partial or graduated steps would almost certainly recom-mend themselves if a government took the problem seriously. If at midnight a president or a premier considers war a significant likelihood within the next twenty-four hours, can he let everybody go to work the next morning? Or should he declare a holiday, so that families stay together, urban commuter transportation is not fouled up, people can stay tuned in to civil defense bulletins, last-minute instructions can be communicated, and some kind of discipline maintained? If the possibility of general war rises above some threshold, perhaps because a vigorous war is in process in some theater, might not the aged and infirm and those distant from shelter facilities be sheltered or readied for shelter; and should not some of the less essential economic functions be shut down? Can a president or a premier leave the entire population in its normal pristine vulnerability to attack, knowing that war has become a significant likelihood? There is the possibility that any sheltering would be a dramatic signal that war was imminent, and would tip the scales toward war itself, and should be avoided. Equally compelling, though, is the notion that sheltering is less dramatic, less dangerously demonstrative, if it can be graduated in a crisis, so that there is no sudden all-or-none shutdown of activity and rush to the shelters.

Sheltering is not the only "passive defense" activity that might be involved. One type of defense against thermal radiation from nuclear weapons—and it is semantically unclear whether this is a passive defense or an active one—is smoke or fog injected into the atmosphere. A thick layer of smoke can make a difference, especially if anti-missile defenses could oblige the enemy to detonate his weapons at a distance. But a smoke layer could not be produced instantaneously after enemy weapons came in sight; it would work best if the smudge-pots were put into operation before the war started. This means that it is most effective if subject to "mobilization," with the attendant danger that it signals something to the other side.

People in shelters cannot stay forever. The usual calculations of how long people should be able to stay in shelters—what the supply of rations should be, for example—relate to how long it might take radioactivity to decay, and cleanup procedures to dispose of fallout, so that the outside environment would be safe. But if we must envisage sheltering as a mobilization step, as something that occurs before war is a certainty, then the endurance of people in shelters is pertinent to the crisis itself. They may well have been in their shelters for two or three weeks without any war having started; and, like aircraft in the air, they coerce the nation's leaders into decisions that reflect the inability of the country to sustain its readiness indefinitely. Of all the reasons for having people able to stay in shelters for an extended period, one of the most important would be to avoid any need to have a war quickly because the people couldn't stand the suspense or the privation any longer.[7]

De-sheltering would be a significant activity. It would be a dramatic signal either that a nation's readiness was exhausted or that the crisis was becoming less dangerous. It would be at least as significant as a withdrawal of troops or diminished alert for strategic forces. In fact, if populations were sheltered, negotiations would concern not only what the crisis was originally about but also the crisis itself. The imminence of war would be at least as important as the originating cause of the crisis, and perhaps dominate negotiations. It is likely that a condition for de-sheltering one's own populations would be the enemy's assuming comparable vulnerability for its own population, whether through synchronized de-sheltering or the enemy's de-sheltering as a condition for our own.

These are not purely hypothetical possibilities; the fact that the United States has only a rudimentary civil defense program does not make these considerations irrelevant. We undoubtedly have in this country a tremendous potential for civil defense in a crisis. If reasonably organized, the labor force and the equipment of the United States might create a good deal of civil defense within a week or a day. There were at least some people who stayed home during the Cuban crisis. That was a mild crisis; but it might have gone differently. If most Americans decided, or were advised, that war was an imminent possibility, they would undoubtedly provide themselves a good deal of protection if they were decently instructed. They could do even better if plans for such a "crash civil-defense program" were available in advance, and if any critical supplies and equipment were pre-positioned for such an emergency. In fact, simply to avoid panic it could be essential to get the population busily at work

7. In a prolonged crisis, sheltered people could take fresh air nearby, perhaps by rotation, and separated families could be reunited; stocking of supplies could continue and emergency measures be taken outside shelters. This possibility eases the hardships of shelter, but complicates planning—unless it goes ignored in the planning.

on civil defense in a crisis, whether filling cans with water, shoveling dirt against fire hazard, educating themselves by television, or evacuating particular areas before panic set in.

Some of the "mobilization" steps might be more dramatic, more difficult, even more important in the absence of prepared civil defense facilities. So the lack of a systematic program would not necessarily mean that the President had no decisions to make, in a crisis, with respect to the population and the economy. It might only mean that he had less cognizance of his options, less control over his own choice, and less knowledge of the consequences for lack of plans and preparations.

So we do have "mobilization procedures" that could become dramatically important in a crisis. They are anomalously called "passive" defenses when they are potentially more "active" than any others. They are not part of our military organization and our weaponry, so we typically ignore them in discussions of our military posture. But there they are, and they could make the brink of war as busy and complicated and frantic as the mobilizations of 1914. We can hope they would not make it as irreversible.

The special danger is that the way these processes work will not be understood before they are put to test in a real emergency. The dynamics of readiness—of alert and mobilization both military and civilian—involve decisions at the highest level of government, a level so high as to be out of the hands of experts. "The bland ignorance among national leaders," writes Michael Howard in describing the mobilization of 1914, "of the simple mechanics of the system on which they relied for the preservation of national security would astonish us rather more if so many horrifying parallels did not come to light whenever British politicians give their views about defense policy today."[8] Being an Englishman, he modestly confined his comment to his own kind. I have no knowledge of how profound the Russian ignorance is of these matters; the American ignorance is surely not "bland," but it must be great. There are only twenty-four hours in the day; and no President, Secretary, Chief

8. "Lest We Forget," p. 65.

of Staff, or national security advisor is likely to master the diplomacy of military alert and mobilization, particularly when it depends on knowledge of how the Soviet machine works, a knowledge that the best intelligence cannot provide us if the Soviet leaders do not understand it themselves. There are only twenty-four hours in their day, too. In managing nations on the brink of war, every decision-maker would be inexperienced. That cannot be helped. Thinking about it in advance can and should make an enormous difference; but it did not in 1914. The only people who thought about it were the people responsible for victory if war should occur, not the people responsible for whether war should occur.[9]

The Problem of Stability in an Armed World

These two modes of potential instability—one arising in the advantage that may attach to speed, initiative, and surprise at the outbreak of war, the other arising in the possible tendency for alert and mobilization procedures to become irreversible, to impose pressure of time on decisions, or themselves to raise the premium on haste and initiative—are undoubtedly the main sources of mischief that reside in armaments themselves. Deliberate war can of course be undertaken, and sometimes credibly threatened, no matter how much stability resides in the weapons themselves; but the extent to which armaments themselves may bring about a war that was undesired, a war that could bring no gain to either side and was responsive to no political necessity, must be closely related to one or both of these two kinds of instability. And it is the *character* of weapons as much as their quantity, probably more than their quantity, that makes the military environment stable or unstable. The character of military forces is partly determined by geography, partly by the way

9. As background for interpreting the events of 1914 and the ensuing war, and even more as background for today's problems, the first two chapters of Brodie, *Strategy in the Missile Age,* are a merciless examination of the way high officials, civilian and military alike, are tempted to evade the awful responsibility for managing military force when things go wrong.

technology unfolds over time, partly by conscious choices in the design and deployment of military force.

If all nations were self-sufficient islands with the pre-nuclear military technology of World War II, mutual deterrence could be quite stable; even a nation that had determined on war would not care to initiate it.[10] With thermonuclear technology the danger of preemptive instability becomes a grave one; weapons themselves may be vulnerable to sudden long-distance attack unless they are deliberately designed and expensively designed to present less of a surprise-attack target. This in turn can imply a choice between weapons comparatively good for launching sudden attack and weapons comparatively good for *surviving* sudden attack and striking back. The Polaris submarine, for example, is comparatively good at surviving attack and striking second; the Polaris missile itself may be good for starting a war, but not compared with its ability for surviving attack. It is an

10. This is meant to be a factual statement and therefore could be wrong. It could be wrong either about the facts or about the way people would perceive the facts. If amphibious assault looks promising because coastal defense or submarine interdiction is underestimated, the mutual deterrence will not be stable even though it ought to be. And if a country exaggerates the security its oceans give it, as the United States may have done up to 1914, it may not take the steps that, together with its oceanic isolation, could give it security. Hudson Maxim estimated in 1914 that, though the United States had great potential for self defense, there were actually three or four countries that could use our oceans as avenues and successfully invade us. He doubted the United States would arm itself until after it had been badly defeated in a war, and he concluded, discouraged, that "Our business at the present time is to pick our conquerors. I choose England." *Defenseless America* (New York, Hearst International Library, 1915), pp. xx, 72–78, 99–108, 120–25. T. H. Thomas, in a most interesting article on "Armies and the Railway Revolution," says that, "One of the most popular anticipations throughout Germany in the early 1840's was that the coming railway network would establish a decisive handicap against offensive wars, and in particular would make impossible a French invasion of German territory. . . . The first actual test of war quite shattered this picture. In the Italian war of 1859, even with incomplete and very imperfect railway systems, large armies were carried rapidly from distant regions to the chosen front of attack, and Napoleon III could launch a major offensive with a speed the first Napoleon could never have attempted." *War as a Social Institution,* Jesse D. Clarkson and Thomas C. Cochran, eds. (New York, Columbia University Press, 1941), pp. 88–89.

expensive weapon compared with other missiles, and the expense goes into making it less vulnerable to attack, not into making it a better weapon for launching sudden attack. To put the same point differently: a reliable ability to strike back with 500 Polaris missiles, after absorbing an attack, corresponds to a first-strike capability of about 500 missiles, whereas a reliable capability to strike back with 500 more vulnerable weapons would require having a multiple of that number, in order that 500 survive attack, and the first-strike capability would be correspondingly larger. To say that the Polaris system provides, for any given level of retaliatory capability, a comparatively small first-strike capability is only to say that it provides, for any given level of first-strike capability, a comparatively large second-strike capability.

If both sides have weapons that need not go first to avoid their own destruction, so that neither side can gain great advantage in jumping the gun and each is aware that the other cannot, it will be a good deal harder to get a war started. Both sides can afford the rule: When in doubt, wait. In Colonel Maude's day, the recommended rule was: When in doubt, act. Act quickly; and if tempted to hesitate, remember that your enemy will not.

The problem does not arise only at the level of thermonuclear warfare. The Israeli army consists largely of a mobilizable reserve. The reserve is so large that, once it is mobilized, the country cannot sustain readiness indefinitely; most of the able-bodied labor force becomes mobilized. The frontier is close, the ground is hard, and the weather is clear most of the year; speed and surprise can make the difference between an enemy's finding a small Israeli army or a large one to oppose him if he attacked. Preparations for attack would confront Israel with a choice of mobilizing or not and, once mobilized, with a choice of striking before enemy forces were assembled or waiting and negotiating, to see if the mobilization on both sides could be reversed and the temptation to strike quickly dampened.

At the thermonuclear level, the problem of preemptive instability appeared a good deal closer to solution in the middle of the 1960s than it had at the beginning of that decade. This was largely due to the deliberate design and deployment of less vulnerable offensive weapons, partly due to a more explicit official recognition of the problem, and perhaps somewhat due to a growing understanding between the United States and the Soviet Union about the need, and some of the means, for avoiding false alarms and avoiding responses that would aggravate suspicion. During the Cuban missile crisis the Soviet Union apparently abstained from any drastic alert and mobilization procedures, possibly as a deliberate policy to avoid aggravating the crisis. The establishment of a "hot line" between Washington and Moscow was at least a ceremony that acknowledged the problem and expressed an intent to take it seriously.

But the problem of instability does not necessarily stay solved. It may be kept solved, but only by conscious efforts to keep it solved. New weapon systems would not automatically preserve such stability as had been attained by the second half of the 1960s. Ballistic missile defenses, if installed on a large scale by the United States or the Soviet Union, might preserve or destroy stability according to whether they increased or decreased the advantage to either side of striking first; that, in turn, would depend on how much better they worked against an enemy missile force that had already been disrupted by a surprise attack. It would also depend on whether ballistic missile defenses worked best in protecting missile forces from being destroyed or best in protecting cities against retaliation. And it would depend on whether ballistic missile defenses induced such a change in the character of missiles themselves, or such a shift to other types of offensive weapons—larger missiles, low flying aircraft, weapons in orbit—as to aggravate the urgency of quick action in a crisis and the temptation to strike first.

Stability, of course, is not the only thing a country seeks in its military forces. In fact a case can be made that some instability can induce prudence in military affairs. If there were *no* danger of crises getting out of hand, or of small wars blowing up into large ones, the inhibition on small wars and other disruptive events might be less. The fear of "accidental war"—of an unpremeditated war, one that arises out of aggravated misunderstandings, false alarms, menacing alert postures, and a recognized urgency of striking quickly in the event of war—may tend to police the world against overt disturbances and adventures. A canoe can be safer than a rowboat if it induces more caution in the passengers, particularly if they are otherwise inclined to squabble and fight among themselves. Still, the danger is almost bound to be too little stability, not too much of it; and we can hope for technological developments that make the military environment more stable, not less, and urge weapon choices on both sides that minimize instability.

The Problem of Stability in a Disarmed World

Much of the interest in arms control among people concerned with military policy became focused in the early 1960s on the stability of mutual deterrence. Many writers on arms control were more concerned about the character of strategic weapons than the quantity, and where quantity was concerned their overriding interest was the effect of the number of weapons on the incentives to initiate war, rather than on the extent of destruction if war should ensue. A fairly sharp distinction came to be drawn between "arms control" and "disarmament." The former seeks to reshape military incentives and capabilities with a view to stabilizing mutual deterrence; the latter, it is alleged, eliminates military incentives and capabilities.

But the success of either depends on mutual deterrence and on the stability of that deterrence. Military stability is just as crucial in relations between unarmed countries as between armed ones. Short of universal brain surgery, nothing can erase the memory of weapons and how to build them. If "total disarmament" could make war unlikely, it would have to be by reducing incentives. It could not eliminate the potential. The most primitive war could be modernized by rearmament, once it got started.

If war breaks out a nation can rearm, unless its capacity to rearm is destroyed at the outset and kept destroyed by enemy military action. By the standards of 1944, the United States was fairly near to total disarmament when World War II broke out. Virtually all munitions later expended by the United States forces were nonexistent in September 1939. "Disarmament" did not preclude U.S. participation; it merely slowed it down.

As we eliminate weapons, warning systems, vehicles, and bases, we change the standards of military effectiveness. Airplanes count more if missiles are banned, complex airplanes are needed less if complex defenses are banned. Since weapons themselves are the most urgent targets in war, to eliminate a weapon eliminates a target and changes the requirements for attack. A country may indeed be safer if it is defenseless, or without means of retaliation, on condition its potential enemies are equally disarmed; but if so it is not because it is physically safe from attack. Security would depend on its being able to mobilize defenses, or means of retaliation, faster than an enemy could mobilize the means to overcome it, and on the enemy's knowing it.

The difficulty cannot be avoided by banning weapons of attack and keeping those of defense. If, again, nations were islands, coastal artillery would seem useless for aggression and a valuable safeguard against war and the fear of war. But most are not. And in the present era "defensive" weapons often embody equipment or technology that is superbly useful in attack and invasion. Moreover, a prerequisite of successful attack is some ability to defend against retaliation or counterattack; in a disarmed world, whatever lessens the scale of retaliation reduces the risk a nation runs in starting war. Defenses against retaliation are close substitutes for offensive power.

Disarmament would not preclude the eruption of crisis; war and rearmament could seem imminent. Even without possessing complex weapons, a nation might consider initiating war with whatever resources it had, on grounds that delay would allow an enemy to strike or to mobilize first. If a nation believed its opponent might rush to rearm to achieve military preponderance, it might consider "preventive war" to forestall its opponent's dominance. Or, if confidence in the maintenance of disarmament were low and if war later, under worse conditions, seemed at all likely, there could be motives for "preventive ultimatum," or for winning a short war through coercion with illicitly retained nuclear weapons, or for using force to impose a more durable disarmament arrangement. As with highly armed countries, the decision to attack might be made reluctantly, motivated not toward profit or victory but by the danger in not seizing the initiative. Motives to undertake preventive or preemptive war might be as powerful under disarmament as with today's weapons, or even stronger.

In a disarmed world, as now, the objective would probably be to destroy the enemy's ability to bring war into one's homeland, and to "win" sufficiently to prevent his subsequent buildup as a military menace. The urgent targets would be the enemy's available weapons of mass destruction (if any), his means of delivery, his equipment that could be quickly converted for strategic use, and the components, standby facilities, and cadres from which he could assemble a capability for strategic warfare. If both sides had nuclear weapons, either by violating the agreement or because the disarmament agreement permitted it, stability would depend on whether the attacker, improvising a delivery capability, could forestall the assembly or improvisation of the victim's retaliatory vehicles or his nuclear stockpile. This would depend on the technology of "disarmed" warfare, and on how well each side planned its "disarmed" retaliatory potential.

If an aggressor had nuclear weapons but the victim did not, the latter's response would depend on how rapidly production could be resumed, on how vulnerable the productive facilities were to enemy action, and whether the prospect of interim nuclear damage would coerce the victim into surrender.

In the event that neither side had nuclear weapons, asymmetrical lead times in nuclear rearmament could be decisive. Whether it took days or months, the side that believed it could be first to acquire a few dozen megatons through a crash program of rearmament would expect to dominate its opponent.

This advantage would be greatest if nuclear facilities themselves were vulnerable to nuclear bombardment; the first few weapons produced would be used to spoil the opponent's nuclear rearmament. Even if facilities were deep under the ground, well disguised or highly dispersed, a small difference in the time needed to acquire a few score megatons might make the war unendurable for the side that was behind. It might not be essential to possess nuclear weapons in order to destroy nuclear facilities. High explosives, commandos, or saboteurs could be effective. "Strategic warfare" might reach a purity not known in this century: like the king in chess, nuclear facilities would be the overriding objective. Their protection would have absolute claim on defense. In such a war the object would be to preserve one's mobilization base and to destroy the enemy's. To win a war would not require overcoming the enemy's defenses—just winning the rearmament race.

Such a war might be less destructive than war under present conditions, not primarily because disarmament had reduced the attacker's capability for destruction but because, with the victim unable to respond, the attacker could adopt a more measured pace that allowed time to negotiate a ceasefire before he had reduced his victim to rubble. Victory, of course, might be achieved without violence; if one side appeared to have an advantage so convincingly decisive as to make the outcome of mobilization and war inevitable, it might then deliver not weapons but an ultimatum.

An International Military Autho ity

Some kind of international authority is generally proposed as part of an agreement on total disarmament. If militarily superior to any combination of national forces, an international force implies (or is) some form of world government. To call such an arrangement "disarmament" is about as oblique as to call the Constitution of the United States "a Treaty for Uniform Currency and Interstate Commerce." The authors of the Federalist Papers were under no illusion as to the far-reaching character of the institution they were discussing, and we should not be either.

One concept deserves mention in passing: that the projected police force should aim to control persons rather than nations. Its weapons would be squad cars, tear gas, and pistols; its intelligence system would be phone taps, lie detectors, and detectives; its mission would be to arrest people, not to threaten war on governments. Here, however, we shall concentrate on the concept of an International Force to police nations—and all nations, not just small ones. The most intriguing questions are those that relate to the Force's technique or strategy for deterring and containing the former nuclear powers.

The mission of the Force would be to police the world against war and rearmament. It might be authorized only to stop war; but some kinds of rearmament would be clear signals of war, obliging the Force to take action. There might be, explicitly or implicitly, a distinction between the kinds of rearmament that call for intervention and the kinds that are not hostile.

The operations of the Force raise a number of questions. Should it try to contain aggression locally, or to invade the aggressor countries (or all parties to the conflict) and to disable them militarily? Should it use long-range strategic weapons to disable the country militarily? Should it rely on the threat of massive punitive retaliation? Should it use the threat or, if necessary, the practice of limited nuclear reprisal as a coercive technique? In the case of rearmament, the choices would include invasion or threats of invasion, strategic warfare, reprisal or the threat of reprisal; "containment" could not forestall rearmament unless the country were vulnerable to blockade.

Is the Force intended to do the job itself or to head a worldwide alliance against transgressors? In case of aggression, is the victim to participate in his own defense? If the Indians take Tibet, or the Chinese encourage armed homesteading in Siberia, the Force would have to possess great manpower unless it was prepared to rely on nuclear weapons. A force could not be maintained on a scale sufficient to "contain" such excursions by a nation with a large population unless it relied on the sudden mobilization of the rest of the world or on superior weaponry —nuclear weapons if the defense is to be confined to the area of incursion. But the use of such weapons to defend, for example, Southeast Asia against neighboring infiltrators, Western Europe against the Soviet bloc, East Germany against West Germany or Cuba against the United States, would be subject to the ordinary difficulties of employing nuclear weapons in populated areas. A country threatened by invasion might rather capitulate than be defended in that fashion. Moreover, the Force might require logistical facilities, infrastructure, and occasional large-scale maneuvers in areas where it expects to be called upon. Keeping large forces stationed permanently along the Iron Curtain is a possibility but not one that brings with it all the psychological benefits hoped for from disarmament.

A sizable intervention of the Force between major powers is not, of course, something to be expected often in a disarmed world. Nevertheless, if the Force is conceived of as superseding Soviet and American reliance on their own nuclear capabilities, it needs to have some plausible capability to meet large-scale aggression; if it hasn't, the major powers may still be deterred, but it is not the Force that deters them.

A capability for massive or measured nuclear punishment is probably the easiest attribute with which to equip the Force. But it is not evident that the Force could solve the problems of "credibility" or of collective decision any better than can the United States alone or NATO collectively at the present time. This does not mean that it could not solve them—just that they are not automatically solved when a treaty is signed. If the Force is itself stateless, it may have no "homeland" against which counter-reprisal could be threatened by a transgressor na-tion; but if it is at all civilized, it will not be wholly immune to the counter-deterrent threats of a transgressor to create civil damage in other countries. These could be either explicit threats of reprisal or implicit threats of civil destruction collateral to the bombardment of the Force's own mobilization base. (The Force presumably produces or procures its weaponry in the industrial nations, and cannot be entirely housed in Antarctica, on the high seas, or in outer space.)

If it should appear technically impossible to police the complete elimination of nuclear weapons, then we should have to assume that at least minimal stockpiles had been retained by the major powers. In that case, the Force might not be a great deal more than one additional deterrent force; it would not enjoy the military monopoly generally envisaged.

One concept needs to be disposed of—that the Force should be strong enough to defeat a coalition of aggressors but not so strong as to impose its will against universal opposition. Even if the world had only the weapons of Napoleon, the attempt to calculate such a delicate power balance would seem impossible. With concepts like preemption, retaliation, and nuclear blackmail, any arithmetical solution is out of the question.

The knottiest strategic problem for an International Force would be to halt the unilateral rearmament of a major country. The credibility of its threat to employ nuclear weapons whenever some country renounces the agreement and begins to rearm itself would seem to be very low indeed.

The kind of rearmament would make a difference. If a major country openly arrived at a political decision to abandon the agreement and to recover the security it felt it had lost by starting to build a merely retaliatory capability and sizable home-defense forces, it is hard to envisage a civilized International Force using weapons of mass destruction on a large scale to stop it. Limited nuclear reprisals might be undertaken in an effort to discourage the transgressor from his purpose. But unless the rearmament program is accompanied by some overt aggressive moves, perhaps in limited war, the cool and restrained introduction of nuclear or other unconventional weapons into the country's population centers does not seem plausible, unless nonlethal chemical or biological weapons could be used.

Invasion might offer a more plausible sanction, perhaps with paratroops armed with small nuclear weapons for their own defense; their objective would be to paralyze the transgressor's government and mobilization. But if this should be considered the most feasible technique for preventing rearmament, we have to consider two implications. We have provided the Force a bloodless way of taking over national governments. And a preemptive invasion of this kind might require the Force to act with a speed and secrecy inconsistent with political safeguards.

There is also the question of what kinds of rearmament or political activity leading to rearmament should precipitate occupation by the Force. In our country, could the Republicans or Democrats campaign on a rearmament platform, go to the polls and win, wait to be inaugurated, denounce the agreement, and begin orderly rearmament? If the Force intervenes, should it do so after rearmament is begun, or after a party has introduced a rearmament resolution in Congress? The illustration suggests that one function of the Force, or the political body behind it, would be to attempt first to negotiate with a potential rearming country rather than to intervene abruptly at some point in these developments.

Again, the character of rearmament would make a difference. Suppose the President presented a well-designed plan to build an obviously second-strike retaliatory force of poor preemptive capability against either the International Force or other countries, but relatively secure from attack. If he justified it on the grounds that the current military environment was susceptible to sudden overturn by technological developments, political upheavals, irrepressible international antagonism, the impotence of the Force for decisive intervention, the corruption or subversion of the Force, or other such reasons, then the authorization of a drastic intervention by the Force in the United States would be less likely than if the President ordered a crash program to assemble nuclear weapons, trained crews, and long-range aircraft. It would make a considerable difference, too, whether rearmament occurred at a time of crisis, perhaps with a war going on, or in calmer times.

The point of all this is simply that even an international military authority with an acknowledged sole right in the possession of major weapons will have strategic problems that are not easy.[11] This is, of course, aside from the even more severe problems of political control of the "executive branch" and "military establishment" of the world governing body. If we hope to turn all our international disputes over to a formal procedure of adjudication and to rely on an international military bureaucracy to enforce decisions, we are simply longing for government without politics. We are hoping for the luxury, which most of us enjoy municipally, of turning over our dirtiest jobs—especially those that require strong nerves—to some specialized employees. That works fairly well for burglary, but not so well for school integration, general strikes, or Algerian independence. We may achieve it if we create a sufficiently potent and despotic ruling force; but then some of us would have to turn around and start plotting civil war, and the Force's strategic problems would be only beginning.[12]

Designing Disarmament for Stability

A stable military environment, in other words, would not result automatically from a ban on weapons and facilities to make

11. Max Lerner, in the book cited earlier, exemplifies the common tendency to confuse the solution of a problem with its replacement by another, in his haste to make the case for drastic disarmament. "If there were an outlawry of aggressive war in any form, enforced by an international authority, a good deal of what is dangerous about total disarmament could be remedied" (pp. 259–60). But so would it if the outlawry were enforced by the United States, the NATO alliance, or the fear of God; and if such outlawry of "aggressive war" could be enforced by a potent *and* decisive *and* credible authority (immune to the dangers of an opponent's "irrationality" that Lerner, as mentioned earlier, thinks may spoil deterrence), we might settle equally well for something more modest, less unsettling, than "total disarmament." Who cares about arms, much, if we can reliably rule out all modes of aggressive warfare (and self-defensive, preventive, inadvertent, or mischievous warfare)? It may be easier for some "authority" to manage its job if all its opponents are "totally disarmed" but this depends on analysis, not assertion. One cannot disagree with Lerner, only question whether he had said anything.

12. For a more extensive, and somewhat more constructive if equally discouraging, treatment of the "Strategic Problems of an International Armed Force," see the author's article by that title in *International Organization, 17* (1963), 465–85, reprinted in Lincoln P. Bloomfield, ed., *International Military Forces* (Boston, Little, Brown, 1964).

them. War, even nuclear war, remains possible no matter how much it is slowed down by the need to mobilize and even to produce the weapons.[13] The two modes of instability that worry armed countries now (or ought to worry them) would be just as pertinent for disarmed countries. The timing of war and rearmament, and the role of speed and initiative, would remain critically important in a world in which the pace of war was initially slowed for lack of modern weapons. There would remain, even in the design of "total disarmament," the difficult choice between minimizing war's destructiveness and minimizing its likelihood. If disarmament is to discourage the initiation of war, to remove the incentives toward preemptive and preventive war, and to remove the danger of unstable mobilization races, it has to be designed to do that. Disarmament does not eliminate military potential; it changes it.

The essential requirement is for some stable situation of "rearmament parity." If disarmament is to be durable, it must be so designed that the disadvantages of being behind in case an arms race should resume are not too great and so that, in the face of ambiguous evidence of clandestine rearmament or overt evidence of imminent rearmament, nations can react without haste. The straightforward elimination of so-called "military production facilities" might, by sheer coincidence, provide the stability; but stability is more likely if there is a deliberately designed system of "stable equal readiness for rearmament." It is impossible to eliminate the ability to rearm; one can only hope to stretch the time required to reach, from the word "go," any specified level of rearmament, and try to make defensive or retaliatory rearmament easier than offensive or preemptive rearmament. One can try to take the profit out of being ahead and the penalty out of being slow and to minimize the urge of either side, in a renewed arms race, to consolidate its advantage (or to minimize its disadvantage) by launching war itself.

It is not certain that maximizing the time required to rearm is a way to deter it. Lengthening the race course does not neces-

13. This was proved in World War II when the United States not only produced nuclear weapons while the war was on but invented them! Next time it would be easier.

sarily lessen the incentive to be first under the wire. It may, however, reduce the advantage of a small headstart; it may allow time to renegotiate before the race has too much momentum; and it may reduce the confidence of a fast starter that he could win if he called for a race.

The likelihood of war, then, or of a rearmament race that could lead to war, depends on the character of the disarmament. If mobilization potentials are such that a head start is not decisive and the race course is long, preemptive action may be delayed until motives are clear. Important elements for stability in a disarmed world would be the dispersal and duplication of standby facilities for rearmament and of reserve personnel or cadres around which rearmament could be mobilized. Dispersal could be important because of the interaction between rearmament and war itself. If a nation could achieve just enough production of weapons to disrupt its opponent's rearmament, it might gain a decisive advantage. Once the race were on, a few easily located facilities for producing nuclear weapons might invite a preventive and very limited war.

The argument here is not that disarmament would be especially unstable, or less stable than the present world of armament. It is that disarmament could be *either* more stable *or* less stable militarily than an armed world, according to how the existing military potential loaded the dice in favor of speed, surprise, and initiative or instead made it safe to wait, safe to be second in resuming an arms race or second in launching attack, and on whether the easiest directions of rearmament tended toward stable or unstable armaments.

It should not be expected that reduced tensions would be the natural consequence of a disarmament agreement, making the existing military potential irrelevant. Not everyone would be confident that disarmament provided a viable military environment or promised the political atmosphere most conducive to peace and good relations. It is hard to believe that any sober person under any conceivable world arrangement could come to believe with confidence that war had at last been banished from human affairs until there had been at the very least some decades of experience. There would be surprises, rumors, and sharp misunderstandings, as well as the usual antagonisms among countries. It is not even out of the question that if something called "general and complete disarmament" were achieved, responsible governments might decide that international apprehensions would be reduced if they possessed more secure, more diversified, and more professionally organized mobilization bases or weapon systems, with more freedom to improve them, drill them, and discuss the strategy of their use. It might be that moderate though expensive modern weapon systems, professionally organized and segregated from the main population centers, would provide less—not more—military interference in everyday life than a "total" disarmament agreement under which every commercial pilot carried emergency mobilization instructions in his briefcase.

Stability, in other words, of the two kinds discussed in this chapter, is relevant to any era and to any level of armament or disarmament. It is just not true that if only disarmament is "total" enough we can forget about deterrence and all that. It would be a mistake to suppose that under "total" disarmament there would be no military potential to be controlled, balanced, or stabilized. If disarmament were to work, it would have to stabilize deterrence. The initiation of war would have to be made unprofitable. It cannot be made impossible.

It is sometimes argued that to perpetuate military deterrence is to settle for a peace based on fear. But the implied contrast be-tween stabilized deterrence and total disarmament is not per-suasive. What would deter rearmament in a disarmed world, or small wars that could escalate into large ones, would be the apprehension of a resumed arms race and war. The extent of the "fear" involved in any arrangement—total disarmament, nego-tiated mutual deterrence, or stable weaponry achieved unilater-ally by conscious design—is a function of confidence. If the consequences of transgression are plainly bad—bad for all par-ties, little dependent on who transgresses first, and not helped by rapid mobilization—we can take the consequences for granted and call it a "balance of prudence."

7

THE DIALOGUE
OF COMPETITIVE ARMAMENT

Nuclear age communications were dramatized by the Soviet-American hot line, a leased transatlantic cable with teletype machinery at both ends. Some people hailed it as a notable innovation; others were simply astonished that, in an age when one can directly dial his mother 3,000 miles away to wish her happy birthday, facilities did not already exist for a more urgent conversation. The hot line is a reminder that even in the era of Telstar and radio-dispatched taxis, facilities for quick communication between heads of government may not exist unless somebody has thought to provide them.

The hot line was foreshadowed in a speech of Secretary Herter's in early 1960. "Observers might prove useful, during a major crisis, helping to verify that neither side was preparing a surprise attack upon the other." And he said that "other arrangements for exchanging information might be developed to assure against potentially dangerous misunderstandings about events in outer space." The possibility that, in a crisis, reciprocal suspicions might be amplified by a feedback process, each side's preparation against surprise looking like preparations for attack, had begun to receive attention by the time of the Geneva negotiations on surprise attack in 1958. Gromyko gave a vivid description at a press conference of "meteors and electronic interferences" causing Soviet aircraft to be launched, in turn causing American bombers to be launched, so that both sides "would draw the natural conclusion that a real attack by the enemy was taking place."

But Gromyko was not the first Russian to be concerned about this feedback. It worried the Czar in July 1914, when he was trying to decide whether mobilization against Austria would alarm the Germans into mobilizing against France and bring on general war. In fact, the germ of the hot-line idea has to be sought still further back. Neither Gromyko nor Herter, nor any modern writer on arms control, has expressed the problem more lucidly than Xenophon did in the fourth century before Christ. Mutual suspicion arose between the Greek army departing Persia and the Persian army that escorted them. The Greek leader called for an interview with the Persian, to try "to put a stop to these suspicions before they ended in open hostility." When they met, he said,

> I observe that you are watching our moves as though we were enemies, and we, noticing this, are watching yours, too. On looking into things, I am unable to find evidence that you are trying to do us any harm, and I am perfectly sure that, as far as we are concerned, we do not even contemplate such a thing; and so I decided to discuss matters with you, to see if we could put an end to this mutual mistrust. I know, too, of cases that have occurred in the past when people sometimes as the result of slanderous information and sometimes merely on the strength of suspicion, have become frightened of each other and then, in their anxiety to strike first before anything is done to them have done irreparable harm to those who neither intended nor even wanted to do them any harm at all. I have come then in the conviction that misunderstandings of this sort can best be ended by personal contact, and I want to make it clear to you that you have no reason to distrust us.[1]

The upshot of this incident is chastening. The "personal contact" so established was used by the Persians to slay the entire leadership of the Greek host; and while we owe to their treachery one of the most rewarding books on strategy in print, we can lament that they did not get arms control off to a more creditable start. The mistake was apparently in thinking that the

1. *The Persian Expedition,* p. 82.

only way to take the danger out of distrust is to replace it with trust.

The hot line is not a great idea, just a good one. It reminds us that arms control need not be exclusively focused on grand schemes to preserve the peace. Actually, the hot line may be largely symbolic. Who could devise a more vivid, simple ceremony to commemorate nuclear age relations than the delivery to the Pentagon of Cyrillic-alphabet teletype machinery, manufactured in the Soviet Union and lend-leased in return for American equipment delivered to the Kremlin. The mere exchange of such facilities probably induces people to think more seriously about communication, so there may be a better basis for knowing what to say, as well as the equipment for saying it, in an emergency.

It is a commentary on the state of our thinking about war and deterrence that the hot line enjoys such novelty. The Republican platform in 1964 singled it out for attention as though it were unnatural, and as though the urgency of possible communications was a sign of intimacy and America's allies should feel dispossessed by this American liaison with the enemy. Journalistic coverage aggravated the novelty by picturing an American President and a Soviet Premier literally on the telephone (as though there were some language they could speak to each other) and even promoted the apprehension that President Kennedy or President Johnson in his pajamas, at three o'clock in the morning, would sleepily give away some remote part of the world without consulting an atlas or the Department of State.

But there is plenty of historical precedent for communication between enemies. Even the world wars were eventually terminated by a process of negotiation that depended on some line of communication that traversed the combat zone and linked enemies in diplomatic contact. If another war should come, especially a big one, time might not permit seeking out a neutral ambassador to serve as go-between, especially if his fallout shelter had no external antenna. Upon reflection almost anyone will agree that the communication that takes place between enemies is the most urgent and that what is "unnatural" in the modern era is the notion that in case of war there could be nothing legitimate for enemies to talk about.

It is hard to imagine any more bitter enmity than that between the Arabs and the Israelis upon the establishment of the State of Israel. Yet during the cease-fire in Jerusalem at the end of 1948 a "hot line" was established—in this case literally a telephone line linking senior commanders on both sides of Jerusalem (English and Arabic being available languages on both sides)—to handle emergencies arising out of the cease-fire arrangements. The idea, I am told, was not dreamed up by civilian arms-control enthusiasts but initiated by the military commanders themselves, who perceived that exchanges of fire and other incidents might need to be handled in a hurry. This was no novelty; Julius Caesar in Gaul, or Xenophon in Persia, understood the crucial importance of communication with the enemy and inflicted the severest penalties on subordinates who did not respect the personal safety of enemy ambassadors.

In an engineering sense, starting a major war is about the most demanding enterprise that a planner can face. In broader strategic terms, *terminating* a major war would be incomparably more challenging. If ever general war should occur there is every likelihood that it would be initiated reluctantly or would occur unintended; getting it stopped in a manner consistent with all that is at stake would be of an importance and a difficulty that eclipsed any other problem that any modern country has ever faced. Some kind of communication would be at the center of the process. Even deciding with whom one is willing to negotiate might be of critical importance. The hot line does not take care of this problem; it only dramatizes it.

The most important measures of arms control are undoubtedly those that limit, contain, and terminate military engagements. Limiting war is at least as important as restraining the arms race, and limiting or terminating a major war is probably more important in determining the extent of destruction than limiting the weapon inventories with which it is waged. There is probably no single measure more critical to the process of arms control than assuring that if war should break out the adversaries are not precluded from communication with each other.

The Continuous Dialogue

A hot line can help to improvise arms control in a crisis; but there is a more pervasive dialogue about arms control all the time between the United States and the Soviet Union. Some of it is unconscious or inadvertent. I have in mind not the formal negotiations that provide headlines from Geneva, but the continuous process by which the U.S.S.R. and the United States interpret each other's intentions and convey their own about the arms race.

The treatment of nuclear weapons is a good example. Nominally there exists a formal limitation on testing; but the inhibitions on nuclear activities surely go far beyond the terms of the treaty, and communication about the role of nuclear weapons has by no means been confined to the formal bargaining about tests. There is an understanding that nuclear weapons are a special category to be differentiated from the more traditional explosives. The emphasis given to conventional forces by the United States over the past several years is based on the notion that, in limiting war, a significant dividing line occurs between conventional and nuclear explosives, that once a nuclear weapon is used in combat, the likelihood of further use goes up. Some kind of communication, formal or informal, deliberate or inadvertent, tends to create, to confirm, or to enhance these expectations. And there has been a good deal of communication about this nuclear-conventional distinction. Singling out nuclear weapons for a test ban itself celebrated a symbolic or psychological difference between nuclear and other weapons. The negotiations helped to put a curse on nuclear weapons and undoubtedly contributed to a class distinction that, if dramatically recognized in peacetime, can hardly be ignored in case of war.

Even denying the difference between nuclear and other weapons may have contributed to this discrimination. Soviet protestations that nuclear weapons would surely be used had a strident and unpersuasive quality, and at least acknowledged

Soviet awareness of where the West would draw a line. Just joining the argument can contribute to the process of discrimination.

There has been similar "communication"—at the Soviet leaders, if not with them—on other arms questions. Military exploitation of space is an illustration. We have indeed made formal proposals for prohibitions on weapons in space, particularly weapons of mass destruction. But the significant communication has been outside Geneva, some of it verbal, some in what we and the Soviet leaders did and did not do. Anyone who read the newspapers and attended to congressional hearings, government press releases, and press conferences probably got the impression that the United States government had no intention of orbiting nuclear weapons, hoped the Soviet Union would not, and would have been obliged to respond vigorously if it became suspected, known, or announced that the Soviet Union had placed nuclear weapons in orbit. A possible reaction, suggested by the history of our reactions to Soviet behavior, would have been to imitate their performance. Another possibility would have been to interfere with their weapon satellites. And a possible reaction, suggested by both Sputnik and the Korean War, would have been to step up the pace of our entire defense program, especially its strategic component, and most especially our space military activities.

That is probably the impression the Soviet leaders got. Again, it is difficult to tell whether the United States government was consciously signaling its position to the Soviet leaders, giving them hints of what to expect if they did and what to expect if they did not orbit weapons. Much of what a government—any government, but especially the American government—says on a sub-ject like this is in response to immediate questions raised by the press and by Congress. Much of what it says is caught up in the momentum of space programs and military programs. There are many important audiences at home and abroad, and the government speaks with many voices. So it would usually be wrong to suppose the existence of a coherent and careful program of communication to any single audience. Nevertheless, one can suppose that behind some of the statements—and undoubtedly behind some of the silences—is an awareness of the official Soviet audience.

The "agreement" between the United States and the Soviet Union about weapons in orbit, embodied in a U.N. resolution that the two countries sponsored in 1963, appeared to be just formal acknowledgment of an understanding that had already arisen outside of formal negotiation. And what more effective way was there for Khrushchev to ratify the understanding that peacetime satellite reconnaissance was now OK (reversing an earlier Soviet position about shooting them down) than to complain to Senator Benton about U-2 flights over Cuba on grounds that satellites were the proper way to accomplish the same result!

Another arena in which signals may have been emitted is city defenses against ballistic missiles. The Soviet leaders announced proudly in the early 1960s that they had solved the "technical problems" of intercepting missiles. Throughout the postwar period the Soviet leaders have put more emphasis on air defense than the United States, and it could be supposed that they had a predisposition toward defensive installations. It looked for a while as though they might try to recover some of their missile-gap stature by being ahead, or claiming to be ahead, in the deployment of ballistic missile defenses, exploiting a breakthrough that might, in the judgment of some people, drastically shift the strategic balance, leapfrog the American missile superiority, and demonstrate Soviet inventive and productive genius.

In the United States some congressmen, some experts, and some journalists seemed to consider ballistic missile defenses the next great step in strategic weaponry. This interest was reinforced by the test ban. Both critics and supporters of the test ban treated ballistic missile defenses as the most significant development that might be inhibited by the suppression of nuclear experiments.

The Administration took the position in the mid 1960s that ballistic missile defenses might or might not prove feasible and economical but that nuclear testing did not appear to be decisive. The implication was that the test ban was not an indirect ban on ballistic missile defenses or on any other major weapon programs. One could also draw the implication that the test ban might have to be reexamined if that judgment on the significance of testing for missile defenses should prove wrong in the light of new developments.

What was communicated to the Soviet leaders in all this? If they read the testimony of defense officials and the journals devoted to space technology, they undoubtedly got the impression that the Administration considered such defenses to be not yet worth procuring but worth an energetic program of development. They could surely suppose that we were not far behind them, and possibly ahead of them, in solving the technical problems and better able than they to afford the cost of a major new dimension of the arms race.

The Soviet leaders may also have noticed that many officials and commentators said that it would be most serious, even disastrous, if the Soviet Union proceeded with a large-scale ballistic missile defense program and the United States did not, and that the United States should compete and keep up in this field even if, judged on their merits, such defenses really did not appear to be worth the cost. The Soviet leaders might recall the spurt to our defense program and our ballistic missiles in particular that was set off by Sputnik and the apprehension of a missile gap. They may have noticed an almost universal opinion that the United States could not afford to be second in advanced military developments of the magnitude of ballistic missile defenses.

It is possible that they caught on, that they came to perceive that a major program of their own (particularly because city defenses could hardly be invisible) would provide motive, stimulation, or excuse in this country for pushing ahead with a comparable development, perhaps at a pace they would find difficult to match. Perhaps they saw that there was a borderline decision yet to be made in this country, and they might tip that decision by rushing ahead with a program of their own or even by exaggerating their progress.

They seemed to quiet down somewhat on the subject. Even a comparison of the original and the revised editions of Marshal Sokolovskii's *Military Strategy* displays some damping of the original confidence and enthusiasm. Probably without intending it the United States may have signaled something to them—something a little like what is called a "deterrent threat" when it applies to foreign adventures, but which in this case applied to the Soviet internal program. It may have been conveyed to them that our reaction to their program would take the profit out of it and only make the arms race more expensive and more vigorous, not only in ballistic missile defenses but in the kinds and numbers of offensive missiles that would have to be procured.

Implicit Bargaining over Arms Levels

With respect to what we call "aggression"—overt penetration of political boundaries with military force—this process of deterrence is taken for granted. But the bargaining process is less explicit and less self-conscious where domestic arms preparations are concerned. We threaten the Soviet Union that if it seeks strategic advantage by invading Turkey or Iran we shall react with military violence; we do not so explicitly threaten that we shall react with military violence if the Soviets seek military advantage through procurement of a large missile or bomber force, or if they seek to deny us an effective force by building missile defenses and bomber defenses. On the whole, we consider war, even a very limited war, an overt act calling for military response; we do not consider arms pre-parations, even when directed against us, an overt provocation requiring or justifying hostilities.

Nevertheless, in principle, an arms buildup with hostile intent might be met with a military response. The concept of preemption suggests that "hostilities" can be initiated by an enemy country within its own borders, entailing quick military response. Mobilization of armed forces has typically been considered nearly equivalent to a declaration of war; at the outbreak of the First World War, "deterrent threats," unfortunately unsuccessful, were aimed at domestic acts of mobilization as well as against overt aggression. And preventive war against an arming opponent has been a lurking possibility at least since the days of Athens and Sparta.[2] More recently, the United States has engaged in directly coercive military threats to deny the Soviet Union the military advantage of advance deployment of missiles. While Cuba is probably best viewed as a political and geographical Soviet move, it can also usefully be viewed as a Soviet effort to achieve quickly and cheaply an offensive military advantage. An interesting question is whether a comparable crash program within the Soviet Union to acquire a first-strike offensive force might be eligible for comparable sanctions.

As a matter of fact, arms-buildup bargaining does seem to take place, though in a less explicit fashion than the overt territorial bargaining that takes the form of alliances, declarations of commitment, and expressions of retaliatory policy. During most of the Eisenhower Administration the American defense budget was a self-imposed restraint on the Western arms buildup. The motivation may well have been mainly economic, but it is a fair judgment that part of the motivation was a desire not to aggravate an arms race. Even when the assumed "missile gap" created grave concern about the vulnerability of American retaliatory forces in 1959 and the Strategic Air Command displayed a lively interest in the rapid enlargement of an airborne alert, the Administration was reluctant to embark on crash military programs, and there was some evidence that its reluctance was a preference not suddenly to rock the arms-race boat.

2. The Corinthian delegates: "You Spartans are the only people in Hellas who wait calmly on events, relying for your defense not on action but on making people think that you will act. You alone do nothing in the early stages to prevent an enemy's expansion; you wait until your enemy has doubled his strength. Certainly you used to have the reputation of being safe and sure enough; now one wonders whether this reputation was deserved. The Persians, as we know ourselves, came from the ends of the earth and got as far as the Peloponnese before you were able to put a proper force into the field to meet them. The Athenians, unlike the Persians, live close to you, yet still you do not appear to notice them; instead of going out to meet them, you prefer to stand still and wait till you are attacked, thus hazarding everything by fighting with opponents who have grown far stronger than they were originally." *The Peloponnesian War,* p. 50.

Moreover, among the many inhibitions on civil defense in this country over the last several years, one was a desire not to add a dimension to the arms race, not to appear frantically concerned about general war, and not to destabilize the defense budget.

There have also been the direct efforts, in disarmament negotiations, to reach understandings about the relation of armed forces on both sides. With the exception of the test ban, these have come to nothing; and the test ban, whatever combination of good and harm it may have done so far, pertinently illustrates the combination of threats and reassurances that, at least implicitly, go with any bargaining process. In addition to the argument, "We won't if you don't," there has been the argument, "And we will if you do."

More visibly and more dramatically, a defense-budget increase was virtually used for display by President Kennedy in the summer of 1961, as a reaction to the Berlin provocations of that year. The alacrity with which Khrushchev responded with announced budget increases of his own that summer made the process look very much like negotiation in pantomime. The fact that Khrushchev was unable or unwilling even to make clear what he would spend the money on, and the fact that many of the increases in American spending were only indirectly related to the Berlin problems that prompted the increase, confirmed the interpretation that these increases were themselves a process of active negotiation, of threats and responses through the medium of the arms race itself.

And when the summit conference in Paris collapsed in May 1960 in the wake of the U-2 incident, Khrushchev showed his sensitivity to this bargaining process. In response to a reporter's question why American forces had gone on some kind of alert the night before, he remarked that it was probably the American Administration's attempt to soften up American taxpayers for a defense-budget increase. In that remark, he showed himself perceptive of the arms-buildup bargaining that goes on between us and alert to the early symptoms of an aggravated arms race.

Communicating Military-Force Goals

I wonder what we communicate about military-force levels. I have particularly in mind the strategic nuclear forces, the medium and long-range bombers and missiles. Verbally, at least, we communicate something, because the most dramatic disarmament proposals in Geneva tend to be concerned with force levels—percentage cuts, freezes, and so forth. Disarmament aside, American force goals must be somewhat related to what number of bombers and missiles we think the Soviet Union has or is going to have; and probably the missile buildup in the Soviet Union is related in some fashion to the size of Western forces. When the Secretary of Defense makes an announcement about the total number of missiles or submarines or long-range bombers this country plans to have on successive dates in the future, he is providing a guideline for Soviet forces planning at the same time.

And presumably Soviet programs, to the extent that we can perceive them with any confidence, have an influence on ours. The Soviet leaders have probably learned that the easiest way to add bargaining power to those in the United States who would like to double our missile force is to enlarge their own, or to seem about to enlarge it, or to find a persuasive way of claiming that it is going to be larger than we had predicted. They probably know that if they display a supersonic heavy bomber with evidence that they are procuring it in significant numbers, the bargaining power in this country of those who want a supersonic bomber will go up. It may go up for good reasons or for bad reasons, but it will go up.

Implicitly, then, if not explicitly, each of us in his own program must influence the other in some fashion. The influence is surely complicated and uneven, indirect and occasionally irrational, and undoubtedly based often on inaccurate projections of each other's programs. But the influence is there. The Soviets may not have realized when they lofted their first Sputnik into orbit that they were doing for American strategic forces what the Korean invasion had done earlier to Western military programs. They might have guessed it; and even if they did not, in retrospect they must be aware that their early achievements in rocketry were a powerful stimulus to American strategic weapon development. The American bomber buildup in the 1950s was a reflection of the expected Soviet bomber forces and air defenses; the "missile gap" of the late 1950s spurred not only research and development in the United States but also weapon procurement. Whether the Soviets got a net gain from making the West believe in the missile gap in the late 1950s may be questionable, but it is beyond question that American bomber and missile forces were enhanced in qualitative performance, and some of them in quantity, by American beliefs.

Here it becomes clear that the so-called "inspection" problem, widely argued in relation to disarmament, is really no more relevant to disarmament than to armament. We always have our "inspection" problem. With or without disarmament agreements we have a serious and urgent need to know as accurately as possible what military preparations the other side is making. Not only for overt political and military responses around the world, but even for our own military programming, we have to know something about the quantity or quality of military forces that oppose us. In deciding whether to plan for 20 or 200 Polaris submarines, for 500 or 5,000 Minutemen, in deciding whether a new bomber aircraft should have special capabilities against particular targets, in reaching decisions on the value and the performance of defenses against ICBMs, in deciding what to include in the payload of a missile we build and how to configure our missile sites, we have to estimate the likely military forces that will confront us year after year throughout the planning period.

We have to use what information we can get, whether from unilateral intelligence or from other sources. If we decide unilaterally to be just as strong, twice as strong, or ten times as strong as the Soviet Union over the next decade, our need to know what the Soviets are doing is as important as if we had a negotiated agreement with them that we should be just as strong, twice as strong, or ten times as strong over the decade.

The difference is apparently that under disarmament agreements it is acknowledged (at least in the West) that each side needs information about what the other is doing. It is even acknowledged that each ought to have an interest in displaying its program to the other in the interest of maintaining the agreement. But this should be equally true without any agreement: the Soviets in the end may actually have suffered from our belief in the missile gap, much in the way they would suffer under a disarmament agreement that provided us insufficient assurance about the pace of their own program. If we insist on a given ratio of superiority and drastically overestimate what the Soviets have, not only do we spend more money but they have to spend more too. They have to try to keep up with us; and in so doing may "justify" ex post facto the program that we had set afoot on the basis of our original exaggerated estimates.

There is undoubtedly, then, some interaction between the forces on both sides. But any actual dialogue is quite inexplicit. There is rarely a public indication of just how American strategic force plans might be adapted to changes in the Soviet posture. The Defense Department does not say that its program for the next several years involves a specified numerical goal that will go up or down by so many hundreds of missiles according to how many the Soviet Union seems to be installing.[3] Nor does it ever appear that an articulate threat of enlargement in American force is beamed at the Russians to deter their own buildup. Any bargaining that the American government does with the Soviet government about force levels is thus quite inarticulate, probably only semiconscious, and of course without any commitments behind it. The Soviet leaders are even less ex-

---

3. There would be precedent for it. Winston Churchill, addressing the House of Commons in 1912 as First Lord of the Admiralty, "laid down clearly, with the assent of the Cabinet, the principles which should govern our naval construction in the next five years, and the standards of strength we should follow in capital ships. This standard was as follows: sixty per cent in dreadnoughts over Germany as long as she adhered to her present declared program, and two keels to one for every additional ship laid down by her." Winston S. Churchill, *The World Crisis 1911–1918* (abr. and rev. ed. London, Macmillan, 1943), pp. 79–80.

plicit, if only because they are a good deal less communicative to the outside world.

Feedback in the Arms Competition

In the short run we can presumably base our military plans on decisions the Soviets have already taken and programs they have already set afoot. There is substantial lead time in the procurement and development of weapons, and for some period, measured in years rather than months, it is probably safe to *estimate* enemy programs rather than to think about *influencing* them. At least, it is probably safe to estimate them rather than to try to influence them in the downward direction. We could probably boost Soviet military production within a year or two, just as they could boost ours by their actions; it is unlikely that either of us would slack off drastically on account of any short-run events—short of a change in regime or the discovery that one's information has been wholly wrong for several years. (The fading of the "missile gap" did not nearly reverse the decisions it had earlier provoked.)

But in thinking about the whole decade ahead—in viewing "the arms race" as an interaction between two sides (actually, among several sides)—we have to take some account of the "feedback" in our military planning. That is, we must suppose that over an appreciable period of years Soviet programs respond to what they perceive to be the "threat" to them, and in turn our programs reflect what we perceive to be that "threat" to us. Thus, by the end of the decade, we may be reacting to Soviet decisions that in turn were reactions to our decisions early in the decade; and vice versa. The Soviets should have realized in 1957 that their military requirements in the middle 1960s would be, to an appreciable extent, a result of their own military programs and military public relations in the late 1950s.

This is the feedback process in principle, but its operation depends on the fidelity of perception and information, biases in the estimating process, lead time in military procurement decisions, and all of the political and bureaucratic influences that are brought to bear by interservice disputes, budgetary disputes, alliance negotiations, and so forth.

An important question is just how sensitive either of us actually is to the other's program. To approach that question, we ought to inquire into the processes by which either of us reacts to the other. These reactions are surely not just the result of a coolly calculated and shrewd projection of the other side's behavior and a coolly calculated response. Nor do the military decisions of either side result simply from rational calculations of an appropriate strategy based on some agreed evaluation of the enemy. Partly they do, but partly they reflect other things.

First, there may be a certain amount of pure imitation and power of suggestion. There is usually a widespread notion that, to excel over an enemy, one has to excel in every dimension. There seems to be a presumption that, if the enemy makes progress in a particular direction, he must know what he is doing; we should make at least equal progress in that direction. This seems to be the case whether in economic warfare, nuclear-powered aircraft, foreign aid, ballistic missile defenses, or disarmament proposals. This particular reaction seems to be based on hunch; it may be a good one, but it is a hunch.

Second, enemy actions may simply remind us of things we have overlooked, or emphasize developments to which we have given too little attention.

Third, enemy performance may have some genuine "intelligence value" in providing information about what can be done. The Soviet Sputnik and some other Soviet space performances may have had some genuine value in persuading Americans that certain capabilities were within reach. The United States' detonation of nuclear weapons in 1945 must have been comparably important in making clear to the Soviets, as to everyone else, that nuclear weapons were more than a theoretical possibility and that it was perfectly feasible to build a weapon that could be transported by airplane.

Fourth, many decisions in government result from bargaining among services or among commands. Soviet performance or Soviet emphasis on a particular development may provide a powerful argument to one party or another in a dispute over weapons or budget allocations.

Fifth, many military decisions are politically motivated, inspired by the interests of particular congressmen or provoked by press comment. Soviet achievements that appear to be a challenge or that put American performance in a poor light may have, beneficially or not, some influence on the political-decision process.[4]

And in all of these processes of influence it is not the true facts but beliefs and opinions based on incomplete evidence that provide the motivating force.

I see no reason to suppose that the Soviets react in a more rational, more coolly deliberate way, than the West. They surely suffer from budgetary inertia, interservice disputes, ideological touchstones, and the intellectual limitations of a political bureaucracy, as well as from plain bad information. Furthermore, both we and the Soviets play to an audience of third countries. Prestige of some sort is often at stake in weapon-development competition; and a third-area public exercises some unorganized influence in determining the particular lines of development that we and the Soviets are motivated to pursue.

On the whole, the evidence does not show that the Soviets understand this interaction process and manipulate it shrewdly. The Korean War, in retrospect, can hardly have served the Soviet interest; it did more than anything else to get the United States engaged in the arms race and to get NATO taken seri-

4. There is even some influence of "fashion" in military interests: scientists who had been wholly uninterested in "big" bombs before the Soviet explosion of a sixty megaton weapon in the Arctic in 1961 appeared to "discover" interesting things about very large weapons, and had more of the facts at their fingertips soon afterward. Ballistic-missile defenses became fashionable in the early 1960s, partly through stimulation by the Soviet Union. The tendency is of course not peculiar to military programs; physical fitness and poverty, like space, show the same phenomenon. Maybe fashion is a good thing if it is reasonably selective; it may be useful to concentrate attention on a few developments, rather than to apportion interest strictly to the merits of programs, especially if interest has to reach "critical mass" before people can concert and communicate. Evidently this is something, though, that there can easily be too much of.

ously. The Soviets may have been under strong temptation to get short-run prestige gains out of their initial space successes; perhaps they lamented the necessity to appeal to a public audience in a fashion that was bound to stimulate the United States. Whatever political gains they got out of the short-lived missile gap which they either created or acquiesced in, it not only stimulated Western strategic programs but possibly gave rise to a reaction that causes the Soviets to be viewed more skeptically at the present time than their accomplishments may actually warrant. Maybe the Soviets were just slow to appreciate the way Americans react; or maybe they, too, are subject to internal pressures that keep them from pursuing an optimal strategy in the arms race. But if on their own they do not understand the extent to which Western programs are a reaction to theirs, perhaps we can teach them.

This kind of thing has happened. Samuel P. Huntington examined a number of qualitative and quantitative arms races during the century since about 1840, and he does find instances in which one power eventually gave up challenging the supremacy of another. "Thus, a twenty-five year sporadic naval race between France and England ended in the middle 1860s when France gave up any serious effort to challenge the 3:2 ratio which England had demonstrated the will and the capacity to maintain." He points out, though, that "in nine out of ten races the slogan of the challenging state is either 'parity' or 'superiority,' only in rare cases does the challenger aim for less than this, for unless equality or superiority is achieved, the arms race is hardly likely to be worthwhile."[5] The latter statement, however, is probably more relevant to a pre-nuclear period, in which military force was for active defense (or overt aggression) rather than for a deterrent based on a retaliatory potential. The British, after all, usually settled for a strong defensive capability, in the form of a navy, and could do this because Britain was an island; land-war technology on the Continent

5. Samuel P. Huntington, "Arms Races: Prerequisites and Results," *Public Policy,* Carl J. Friedrich and Seymour E. Harris, eds. (Cambridge, Harvard University Press, 1958), pp. 57, 64.

could not so overtly discriminate between offensive and defensive force, and anything less than parity meant potential defeat.

It is hard to believe that the Soviet Union could openly acknowledge that it was reconciled to perpetual inferiority. It would be difficult for them even to acknowledge it to themselves. It might, however, be possible to discourage very substantially their genuine expectations about what they could accomplish in the arms race. At the level of strategic weaponry they have, for one reason or another, had to content themselves with inferiority during the entire period since 1945; they may be able to content themselves indefinitely with something less potent, less versatile, less expensive than what the United States procures. In any event they may be led to believe that they cannot achieve a sufficiently good first-strike capability to disarm the United States to a sufficient extent to make it worthwhile.

If Soviet leaders try to discern how our force levels relate to their own, do they see any close association between that relationship and the disarmament proposals we make in Geneva? It is hard to say. For reasons that are not altogether clear, there has been a tendency for disarmament proposals to assume that some kind of parity or equality is the only basis on which two sides could reach an agreement. Disarmament negotiations typically assume also that arms are to go downward, rather than to stop where they are or just to rise less than they otherwise might. (If this were not so—if a freeze had been recognized all along as a drastic measure of arms control—the absurdity of "inspection proportionate to disarmament" could not so easily have been disguised.) But if the Soviet leaders have felt themselves to be engaged in a tacit dialogue on strategic forces, they probably perceived it as one in which the United States was concerned with how much superiority it wanted at the strategic level and where to taper off a rising inventory of missiles. Thus the conscious and articulate dialogue at Geneva and the less conscious, inexplicit dialogue that goes on continually in Washington and Moscow are based on alternative premises—perhaps properly so, the Geneva dialogue having to do with what the West would consider appropriate if the whole basis of military relations between East and West could be formally changed.

If this ongoing dialogue, the one that really counts as far as military planning is concerned, gets little help from the verbal exchanges in Geneva, is it actually hindered by them? Do we get a confused message across to the Soviet leaders, and possibly they to us, because of the noise emanating from Geneva and uncertainty about which is the authentic voice?

I used to worry about this; and it may be that on matters narrowly identifiable as "arms control" the disarmament negotiations are a noisy interference. But I doubt whether they significantly obstruct the Soviet ability to get the message from Washington unless the Soviet leaders are so ill-attuned that they would not get the message anyway. (They may be ill-attuned, as their Cuban overstep suggests, but we cannot blame it on Geneva.) In America we have been suffering from proliferation in recent years—of cigarette brands, not nuclear weapons— and smokers eager to try new brands are usually anxious to discriminate between mentholated and ordinary. As far as I know, there has been no collusion between cigarette manufacturers and their millions of customers on a signal, and there may not have been even among the manufacturers, yet there has arisen a fairly reliable color signal: mentholated cigarettes are to be in green or blue-green packages. I think by now the Soviet leaders have discerned that statements datelined Geneva are mentholated.

Disarmament advocates may not like the idea that any understandings with the Soviet Union on force levels are reached through the process of military planning and a half-conscious, inarticulate dialogue with the enemy, unenforceable when reached, subject to inspection only by unilateral intelligence procedures, and reflecting each side's notion of adequate superiority or tolerable inferiority. Opponents of disarmament may not like the idea that the executive branch or the Defense Department, even inadvertently, may accommodate its goals to Soviet behavior or try to discern and manipulate enemy intentions. But the process is too important to be ignored and too natural to be surprising. Nor is it a new idea.

In 1912 Churchill was chagrined at the naval procurement plans of the Kaiser's government, which was about to purchase a quarter again as many dreadnoughts as Churchill had expected them to. He wondered whether the Germans appreciated that the result of their naval expansion would be a corresponding British expansion, with more money spent, tensions aggravated, and no net gain to either from the competition. The Cabinet sent the Secretary of State for War to Berlin to communicate that if the Germans would hold to their original plan, the British would hold to theirs; otherwise Great Britain would match the Germans two-for-one in additional ships. Churchill thought that if the Germans really did not want war they would be amenable to the suggestion, and that nothing could be lost by trying.

Nothing was lost by trying. In his memoirs, Churchill displays no regrets at having had the idea and having made the attempt. He had not had a "disarmament agreement" in mind; he simply hoped to deter an expensive acceleration of the arms race by communicating what the British reaction would be. He did it with his eyes open and with neither humility nor arrogance.[6] Decades earlier the French had been persuaded of the futility of trying to overtake British naval tonnage, and it made sense to see whether the Germans could accommodate to the same principle. Like the Greek "hot line" to the Persians, it was a good idea. Unlike it, it was undertaken with open eyes and no commitment; the emissary was not assassinated in the enemy's camp nor was naval procurement held up pending its outcome.

Essentially, this process of discouraging the Soviets in the arms race is no different from trying to persuade them that they are getting nowhere by pushing us around in Berlin. In Berlin, as in Cuba, we have tried to teach them a lesson about what might have been called "peaceful coexistence," if the term had not already been discredited by Soviet use. We did, in the Cuban event, engage in a process intended to teach the Soviets something about what to expect of us and to discourage them from making future miscalculations that might be costly for both of

6. *The World Crisis*, pp. 75–81.

us. In the vicinity of Berlin we have been trying, not without success, to persuade them that certain courses of action are doomed to futility. Maybe we could communicate something similar with respect to the arms buildup itself.

It does seem worthwhile to have some design for managing the arms race over the next decade or two. It is prematurely defeatist to suppose that we could never persuade the Soviets, at least tentatively, that this was a race they could not win. The principle of "containment" ought to be applicable to Soviet military preparation. However constrained they are by an ideology that makes it difficult for them to acknowledge that they are bested or contained, they must have some capacity for acceptance of the facts of life. Perhaps the American response can be made to appear to be a fact of life.

This is a kind of "arms control" objective. But it differs from the usual formulation of arms control in several respects. First, it does not begin with the premise that arms agreements with potential enemies are intrinsically obliged to acknowledge some kind of parity. (But since there are many different ways of measuring military potency, it might be possible to permit an inferior power to claim—possibly even to believe in—parity according to certain measures.) Second, it explicitly rests on the notion that arms bargaining involves threats as well as offers.

It may be impolite in disarmament negotiations explicitly to threaten an aggravated arms race as the cost of disagreement. But, of course, the inducement to agree to any reciprocated modification of armaments must be some implicit threat of the consequences of failure to agree. The first step toward inducing a potential enemy to moderate his arms buildup is to persuade him that he has more to lose than to gain by failing to take our reaction into account. (It could even be wise deliberately to plan and to communicate a somewhat excessive military buildup ratio relative to the Soviet force in order to enhance their inducements to moderate their own program. This sort of thing is not unknown in tariff bargaining.)

Of course, some important dimensions of the arms buildup cannot be characterized as an "arms race." A good many mili-

tary facilities and assets are not competitive: facilities to minimize false alarm, facilities to prevent accidental and unauthorized acts that might lead to war, and many other improvements
in reliability that would help to maintain control in peacetime
or even in war. That is to say, it may be no disadvantage to one
side that the other should make progress on those particular
capabilities. Indeed they may be as desirable in an enemy force
as in one's own. We may not react to the particular steps they
take along these lines, but if we do it is not to make up some lead
we have lost. The hardening or dispersal of the missile force
within a fixed budget may represent an "improvement" in a
country's strategic posture but may not be especially deplored
by the other side; the second country may actually react, in its
own planning, in a direction that is reassuring rather than
menacing. A missile-hardening race is not the same as a
missile-numbers race. Getting across to the Soviet Union the
kind of reaction they can expect from us, therefore, involves
more than a quantitative plan; it involves getting across a notion
of the kinds of weapon programs that appear less provocative
and those that would appear more so. The Cuban affair is a
reminder that there can be a difference.

If, in our attempts to plan a decade or more ahead, we take
seriously the problem of arms-race management and consider
the interaction between our programs and the Soviets', we have
to engage in quite a new exercise: thinking about the kind of
military-force posture that we would like the Soviets to adopt.
Typically in discussions of military policy we treat the Soviet
posture either as given or as something to be determined by
factors outside our control, to which we must respond in some
adequate way. As a result, nothing appears to be gained by
thinking about *our* preferences among alternative *Soviet* postures, doctrines, and programs. But if we begin to examine how
we might influence the Soviet posture, we have to consider
which alternative Soviet developments we prefer and which we
would deplore.

Quantitatively, this requires us to decide whether we want a
maximal or a minimal Soviet effort. Qualitatively, it requires us
to consider alternative Soviet weapons systems and force

configurations. The kinds of arguments we occasionally have in this country about first-strike versus second-strike forces, the merits of active and passive defenses of the homeland, a counterforce or a city-destroying general-war doctrine, and a mix of forces between intercontinental and limited-war capability—all of these arguments we can imagine taking place within the Soviet Union, too. If we are to have any influence on the outcome of those arguments, diffuse and indirect though it may be, we have to decide in what direction we want to exert it.

On the Subdual of Violence

Of all the military matters on which the American and Soviet governments communicate, none is more important than how a major war might be conducted if it actually occurred. This issue is important because it may have more to do with the extent of devastation in such a war than the accumulated megatonnages in the forces. And it is a matter to which some kind of communication is essential. Expectations arrived at before the war might be decisive, not only in making the search for viable limits successful but even in making the effort worthwhile and making the governments sensitive to the possibility.

It seems to have been a discovery of recent years that some reciprocated restraints would make sense in a general war, in the unhappy event such a war should occur. Since even this primitive concept had not been obvious, some conscious thought and communication are evidently essential. Of all the reasons for observing restraint in such a war and for going to some trouble in peacetime to see that restraints could in fact be observed, the strongest reason is that it makes sense for the other side, too. But the other side must know it, must be equipped to perceive restraint if it occurs, and must have equipped itself to be discriminating in its own fashion, too.[7]

The first hints of such a policy in the United States appeared early in the Kennedy Administration, in such things as the defense budget message. The first official articulation of it was in

7. See Chapter 5, above.

Secretary McNamara's June 1962 speech in Ann Arbor, Michigan. Some unofficial discussion of the idea had appeared in the literature before then; it had already been attacked with equal vigor by persons identified with peace movements and by persons identified with the most intransigent "hard" military line. Secretary McNamara amplified the signal by several decibels, giving it official expression in a major speech.

The Soviet response to the official American suggestion that even a thermonuclear war, involving strategic weapons and homelands on both sides, might be restrained and brought under control, is still in process. If the idea took that long to appeal to our own government, especially when it was not exactly a new idea, one should not expect it to appeal instantly to Soviet leaders particularly when it emanates from the United States. They may have to think about it, argue about it, analyze its compatibility with their own weapons plans, and find a meaning and interpretation of it that correspond to their own strategic position. There is every reason to suppose that the concrete implications for the Soviet Union of such an idea would be very different from the implications for the United States, if only because the strategic forces of the two sides are both unlike and unequal.

Furthermore, the Soviet leaders may be inhibited from acknowledging wisdom in an idea on which the Defense Department holds a copyright. And they may have committed themselves to some early rebuttals that embarrassed their further treatment of the subject. For the strategically inferior power there is a dilemma to be taken quite seriously: to maximize deterrence by seeming incapable of anything but massive retaliation, or to hedge against the possibility of war by taking restraints and limitations seriously.

On this and other questions about nuclear weapons and nuclear war, there are signs that the dialogue between East and West is becoming more real, more conscious, less like a pair of monologues and more like two-way communication. The second edition of *Military Strategy*—the landmark official Soviet strategic publication, which appeared in 1962 in the Soviet

Union—showed unmistakable signs of response to the Western response, a feedback cycle. According to Thomas W. Wolfe, author of the introduction to one of the American translations of the volume, the Soviet authors were becoming aware that an important audience for their work is in the Western nations and that it matters how they communicate with that audience.[8]

As a matter of fact, the American treatment of that particular book may have set in motion a process of conscious communication that will be as important as anything that goes on in Geneva, perhaps more important, completely eclipsing Pugwash Conferences and other minor efforts to get communication going between East and West on security matters.

The principle that may have been uncovered by the American treatment of Marshal Sokolovskii's book is a simple one. You get somebody's attention much more effectively by listening to him than by talking at him. You may make him much more self-conscious in what he communicates if you show that you are listening carefully and taking it seriously.

Two translations of the book appeared quickly in the United States, both with introductions by noted American scholars of Soviet military affairs. One was a quick translation, known to have been done for the United States government; the other was a translation supervised by three well-known experts on Soviet policy at the RAND Corporation.[9] Columnists gave a good deal of attention to the book. There was every sign that it was being carefully read within the government and by scholars, military commentators, journalists, and even students. No wonder the Soviet authors in their second edition reacted to some of the Western commentary, "corrected" some of the "misconceptions" of their overseas readers, and quietly corrected some of

8. T. W. Wolfe, "Shifts in Soviet Strategic Thinking," *Foreign Affairs, 42* (1964), 475–86. This article has since been incorporated into Wolfe's superb *Soviet Strategy at the Crossroads* (Cambridge, Harvard University Press, 1964).

9. *Soviet Military Strategy,* RAND Corporation Research Study, H. S. Dinerstein, L. Gouré, and T. W. Wolfe, eds. and translators (Englewood Cliffs, Prentice-Hall, 1963), V. D. Sokolovskii, ed. of original Russian edition; also V. D. Sokolovsky, ed., *Military Strategy* (New York, Praeger, 1963), introduction by R. L. Garthoff.

their own text. There are indications that some of the more extreme doctrinal assertions have been softened, as though in fear the West might take them too seriously![10]

This strange, momentous dialogue may illustrate two principles for the kind of noncommittal bargaining we are forever engaged in with the potential enemy. First, don't speak directly at him, but speak seriously to some serious audience and let him overhear. Second, to get his ear, listen.

10. Wolfe gives an example that is almost too good to be true. Four of the Sokolovskii authors, in an article in *Red Star,* took issue with the American editors on whether Soviet doctrine considered "inevitable" the escalation of limited wars into general war. To prove they had never argued for "inevitability," they quoted a passage from their own book—a passage that the American edition had actually reproduced in full—and deleted in quoting themselves the very word "inevitable" from their own quotation! *Foreign Affairs, 42* (1964), 481–82; *Soviet Strategy at the Crossroads,* pp. 123–24.

INDEX